Race and Science

Race and Science

Scientific Challenges to Racism in Modern America

Edited by

Paul Farber and Hamilton Cravens

Oregon State University Press
Corvallis

The paper in this book meets the guidelines for permanence and durability of the Committee on Production Guidelines for Book Longevity of the Council on Library Resources and the minimum requirements of the American National Standard for Permanence of Paper for Printed Library Materials Z39.48-1984.

Library of Congress Cataloging-in-Publication Data
Race and science : scientific challenges to racism in modern America / edited by Paul Farber and Hamilton Cravens.
p. cm.
Includes bibliographical references and index.
ISBN 978-0-87071-576-1 (alk. paper)
1. Science--Social aspects--United States. 2. Science and law--United States. 3. Racism--United States. 4. Science--Social aspects--North America. 5. Science and law--North America. 6. Racism--North America. I. Farber, Paul Lawrence, 1944- II. Cravens, Hamilton.
Q175.52.U5R33 2009
305.800973--dc22
2009014385

Oregon State University Press
121 The Valley Library
Corvallis OR 97331-4501
541-737-3166 • fax 541-737-3170
http://oregonstate.edu/dept/press

Contents

Introduction

Hamilton Cravens

I.

Race might be thought of as the third rail of American history, that line that carries the hottest and most dangerous currents in the lives of Americans—rich, poor, of moderate means, of either gender, and of any ethnicity, religion, and color. The authors of the essays in this book on race in America initially delivered them as papers at Oregon State University in April 2006 at a conference on the general theme of the parallelisms between two sources of authority for thoughts and deeds with regard to race: the sciences and the law. Parallelisms between the two have existed in our nation's past, and to a considerable extent they still do. We are an autonomous American civilization developed in the wake of the American Revolution, and it was an undeniable fact that scientists and lawyers (and beyond the lawyers, politicians and public men, more generally) thought of these two sources of empirical and moral authority as possessing absolute truth and of being positively true. Science and the law were indeed among the original inspirations for philosophical positivism, the notion that truth can be obtained through rational and empirical inquiry. In this and related senses, we understood that we had an important insight into the American culture. For was it not necessary for the respectable, established citizens of this slaveholding republic—with all the tensions, contradictions, and cognitive dissonances implied therein—to believe that the races of mankind were both ordained and fixed by nature and God? Did not the democratic dogma brought forth by the American Revolution, with its ringing phrase from the Declaration of Independence that "all men are created equal," create a problem for white Americans as they contemplated the nonwhites in their midst, the aboriginal indigenous peoples, and the Africans, slave and free? For if the nonwhites among them were not in fact equal to themselves, then the majority whites had to believe that these peoples were inferior by their very nature and did not deserve to be equal with the white population. That was in fact how the cognitive dissonance between social fact and social perception could be—and were—resolved.

Indeed, when we had the conference, and discussions during and after it, further reflection and investigation on our part suggested that we had identified a much larger phenomenon. The major professions of American life—science, the churches, the law, the political and business elites—not to mention many ordinary Americans, were heavily invested in what might be called *racial essentialism*. Racial essentialism might be simply defined as the idea that the key to any individual's quality or lack thereof was his or her racial identity. This was a highly materialistic point of view, but from the 1830s on, Americans believed that what pigeonholed people into various groups (or races) was a common material experience among each particular group. Obviously each group had a common material experience that differed from those of other groups. And indeed it was in the 1830s that the secular notion of the *group*—a class, a race, a nationality, a gender, and the like—took shape in American and European intellectual life on both popular and elite levels for the first time. Racial consciousness on both sides of the Atlantic as a formal schema dividing up the populations of the world was to be expected.[1]

In turn, this was part and parcel of what might be considered the modern point of view, which was itself the product of the development of what the German philosopher Juergen Habermas has called a public sphere of civilized life, which has increasingly characterized American life since the Revolution.[2] It took perhaps a generation beyond the American Revolution for these views to solidify, but it is remarkable to note that positivism—the belief that knowledge about the natural and the human worlds is absolutely true in every regard—antedated the publication and arrival in America of Charles Darwin's *Origin of Species*, published in 1859, by a good decade.[3] Belief in a *secular* absolute truth, then, has been as American as apple pie, an integral part of the American culture and, indeed, of our common Atlantic civilization. After all, those who lived in Western Europe and North America have been the cultural inheritors of the scientific revolution of the seventeenth century and the eighteenth-century Enlightenment, as well as the democratic revolutions of the last several decades of the eighteenth century. Those momentous events enshrined belief in reason, empirical inquiry, and natural law in this Atlantic civilization.[4]

Every paper presented in Corvallis is published here, generally with modest revisions. Professor Paul Farber and I have penned fresh essays to round out the volume. What we have before us is a history of the development, from the larger society, of notions of racial essentialism and of racial superiority and inferiority in the nineteenth and the early twentieth centuries, and how such ideas and proscriptions have fared since. There is a relatively large and sophisticated collection of literature on science and race in nineteenth-century America. We celebrate that fine work and only wish to add that its topic, in many of its ramifications, has been so well discussed that we need no independent essays to document or represent that phenomenon in this volume.[5] We do wish to add—indeed, to emphasize—that although the ideas of science and the actions of men of science contributed mightily to racial essentialism, the concept existed widely and deeply in the culture, and long before that there were "scientific" ideas about nature that suggested racial essentialism, or at least something more than a crude recognition that different groups of persons of color might differ from one another in ineluctable ways.

II.

From the beginning of European settlement of North America, the colonists had to confront the fact that they would have to interact with groups of peoples far different from them. Of course there were the aboriginal Native Americans, in all their diversity of custom, language, religion, technology, and many other aspects of daily existence. But no less strange were the Africans brought over as slaves—men, women, and even children, members of different tribes in their homelands who spoke different languages and practiced unfamiliar customs. What impressed the white European settlers the most about these others was that they were persons of color, and, therefore, of different groups of humanity—of different races, that is. On the European subcontinent, white Europeans had succeeded in keeping their sphere of influence for white Europeans only, although it was clearly peppered by many different nationalities and ethnic groups. The only exception was the Muslim population's occupation of Spain, which ceased before the colonization of North America.

From the beginning, North America was and would be a multiracial society. That was the undeniable reality. And unlike Central and South America, where a relatively small European population managed the indigenous peoples and persons of African descent, free and slave, in North America the settlers sought to create permanent societies of white persons that employed Africans as hewers of wood and haulers of water and pushed the indigenous peoples of their part of the hemisphere farther and farther away from their settlements. The simple truth was that there was no place for the indigenous populations as groups in the white settlements. The facts of demography made the North American racial situation very different from that of the central and southern parts of the Western Hemisphere. It was not that there was no race consciousness in Central and South America—far from it. But in those parts of the New World, there were plenty of persons of darker hues who were not slaves or even necessarily persons of color whose lives were touched, let alone managed, by the small circles of European priests, colonial administrators, and settlers in the colonial era.[6]

Clearly, racial essentialism mattered throughout the New World, and just as clearly, racial oppression and slavery for Africans, and exclusion and removal for Native Americans, were part and parcel of the history of North America from its settlement in the seventeenth century onward.[7] As generations of excellent scholars of the American experience have shown, slavery and race were deeply entwined into American institutions and life.[8] However, this did not exclude the Native American tribes, as the work of many scholars, above all Francis Paul Prucha, has massively demonstrated.[9]

And thus we come to our first three essays, all of which clearly demonstrate, if such evidence were truly required, of how embedded racial essentialism was in the very texture of American life. Professor Edward J. Larson adapts an essay from his most recent book, *A Magnificent Catastrophe*, published in 2007, in which he demonstrates how deeply racial essentialism—the oppression and subjugation of the African American population by the majority white population—was enmeshed into the very fabric of American public life.[10] His subject is the first contested presidential election of 1800, in which the moderate Federalist candidate and incumbent, John Adams of Massachusetts, lost to Thomas Jefferson of Virginia, who represented both the Republican Party and the interests of slaveholders everywhere in the new Republic. Professor Larson navigates the convoluted twists and turns of the politics of the electoral contest with great skill and panache, and readers are directed to the chapter and, ultimately, to the engaging book for the complete story. On this large canvas, race plays an important part, for it helps define the differences among the various political interests involved in that contest. Yet what engages our attention here is Gabriel's revolt that summer.

Who was Gabriel? Born in 1776, Gabriel was a large man with even bigger ideas. A semiliterate slave of African ancestry and a skilled blacksmith, Gabriel often worked for others in the Richmond, Virginia, area; his master rented him out to other employers, and it is doubtless that this circulation among other slaves and free whites heightened his thirst for freedom. He was also a rambunctious, even uppity, person; in a tussle with a white overseer, from whom he stole a pig, he bit off the man's ear and was punished by having his left hand branded. In the spring and summer of 1800, he planned a massive slave revolt in his corner of southeastern Virginia in the vicinity of the state capitol, Richmond. He fashioned weapons (swords from sickles) and spoke with many other slaves—probably too many, as Professor Larson points out. The results were predictable, for the white rulers of the Commonwealth could not afford any challenge to its authority, including, naturally, a servile revolt. There was simply no question of crossing the color line. To be black was to be presumed a slave; there were free African Americans, but Gabriel and his confederates were not among them. They crossed that racial barrier, challenged it, and paid the price for their actions.

Racial essentialism was not merely a notion or practice of white Americans, as Professor Fay A. Yarbrough reminds us in her careful discussion of the development of changing official Cherokee legal notions of race during the century following the American Revolution. Her evidence comes from three distinct legal cases. The tale begins in the later eighteenth century, with the slave woman Molley, who was of African descent, who arrived in the Cherokee Nation just prior to the Revolutionary War as the slave of a white trader who had killed his Cherokee wife. Cherokee law and custom demanded vengeance for the murder of a member of a Cherokee clan, and the trader, Sam Dent, successfully offered his slave, Molley, whom he had bought in neighboring Georgia, as a replacement for his dead wife. The Cherokee town council meeting accepted Molley and all her future descendants as members of the clan and the Cherokee nation. Thus the Cherokee clan emancipated Molley, made her an equal with all other Cherokee, and gave her a new name, Chickaua, thus formally recognizing her emancipation.

By the mid-1820s, however, Cherokee racial attitudes were beginning to change to harden the distinction between Cherokees and African Americans. Shoe Boots, a full-blood Cherokee warrior and war hero, petitioned the Cherokee Council to have the three children he had with his slave partner made full citizens of the Cherokee Nation, which was tricky since Cherokee citizenship was decided by matrilineal descent and clan membership, not race, and because most persons of African descent among the Cherokee were slaves and thus property, not persons eligible for membership in the clan and Nation. The council ratified his children as full Cherokee citizens but clearly disapproved of his continuing sexual relationship with a black slave woman. Forthwith, the council forbade intermarriage between black slaves and Indians or whites. Shoe Boots had two more children with his black slave. After his death in 1829, all his children became the slave property of a white man. Before the 1830s, the Cherokee Council ruled that the children of any Cherokee of African descent could never be citizens, although mixture with whites was thoroughly acceptable.

Between the 1830s and the 1880s, there was much change and tumult for the Cherokee Nation. To a very considerable extent, changing racial attitudes within the Nation were similar to those in American society at large with regard to blacks. And that led to trouble. Removal of the Cherokee people to what is now northeastern Oklahoma in the 1830s, the conflicts that the surviving members of the Nation had over interests and arrangements there, the Civil War, and the emancipation of black slaves in both the United States and the Cherokee Nation as a consequence of that bloody war all led to a new context of race relations in the Cherokee Nation that closely mirrored that in the United States of America. In the antebellum years, most blacks in the Cherokee Nation as well as in the American slave states had been bondsmen. Neither the Cherokees nor white Southerners thought of blacks as equal to themselves or deserving of equality. Hence, as Professor Yarbrough deftly shows us, the two groups were similar in that each had a society in which status depended on racial identity. This was indeed the culmination of racial essentialism among the Cherokee, a people whose racial attitudes paralleled those of the larger white society—and rather surprisingly at that.

An even more extravagant, not to say rancid, example of racial essentialism is provided in Professors John P. Jackson, Jr., and Andrew S. Winston's skillful portrait of Earnest Sevier Cox, the man they dub "the last repatriationist," whose career as a champion of racist ideas spanned much of the twentieth century. From 1920s to his death in the 1960s, Cox was the most well-known white spokesman for the repatriation of African Americans from the United States to their "authentic" homeland, Africa. Here was a version of racial essentialism with a vengeance: whites and blacks were essentially different races, and any admixture—even any geographical proximity—would be a disaster in every sense of the word for both races. He clothed his ideas in a radical scientific racism, which was certainly respectable among some groups of American, British, and German scientists in the post–World War I years, although Professors Jackson and Winston point out that Cox needed no natural science for his racism: history had proof enough of the vast gulf separating the races.

As a young man, Cox began his career as a Methodist minister and then took graduate work in sociology at the University of Chicago, where he wrote a series of papers in which he insisted that African Americans constituted a race that was inferior to the white race. With the encouragement of his teachers, especially W. I. Thomas and Charles Henderson, Cox embarked on a tour of Africa and returned to the United States in 1914 after four years there. Billing himself as an expert in ethnology, he attached himself to Senator James K. Vardaman, a segregationist from Mississippi, while he worked on a book on relations among the races, especially blacks and whites. Vardaman proudly read selections from his protégé's manuscript into the

Congressional Record. Soon Cox became a public figure, giving lectures in scattered venues and insisting that the colored races had never produced a civilization. Unable to find a publisher, he finally paid for the printing of his first book, *White America*, in 1923, which was a restatement of Madison Grant's 1914 book, *The Passing of the Great Race*. With this credential, Cox was able to latch on to the political movement in his home state of Virginia for "racial integrity," meaning no mixture of the races, and in the interwar years, Cox became a champion of repatriation on the premise that when blacks and whites mixed, disaster was the result. He claimed that if all blacks could be relocated in Africa, where they belonged, and all whites could be in North America and Europe, then everything would be right in the world. In the interwar years, Cox became a kind of curious figure. Strictly speaking, he was not a segregationist, let alone an integrationist; he was a separatist. He did appeal, however perversely, to black nationalists such as Marcus Garvey, who despised the white man and his promises and wanted no intercourse whatever with white civilization. At the same time, Cox appealed to pro-Nazis at home and abroad, not to mention hardcore homegrown American racists. Cox was a kind of magnet, attracting all the peculiar strands of antiblack and antiwhite ideologies throughout the land in the interwar years. With the Second World War and the emergence of Nazi Germany as the nation's mortal enemy, Cox had to watch himself, and it was only after that horrific conflict had ended that he resurfaced, this time again as a repatriationist and an advocate of Teutonic, and therefore white, supremacy. Thus he was "rediscovered" by white Southerners, this time in the wake of the United States Supreme Court's 1954 desegregation decision, *Brown v. Board of Education*. Segregation, he insisted, would not save America but in fact lead to race mixing and inevitable decline. In a pamphlet titled *Unending Hate*, which he published in response to the Court's *Brown* decision, he insisted that interracial harmony was an illusion, a nonsensical goal because it would be impossible to achieve. Yet there was a surprise or two with Cox. He was also a consistent anticolonialist. He argued that the white imperialist powers should leave Africa to the Africans. No race mixture meant more than no intermarriage; it meant no geographical proximity. Geographic isolation was the only way to keep the races pure. Crudely put, Cox wanted all African Americans relocated to Africa. How this was supposed to be financed, let alone agreed to by the various interests involved, Cox never bothered to explain; apparently arguments about the correctness of racial essentialism sufficed for him. As William Shakespeare penned Falstaff's famous line, discretion is the better part of valor.

III.

Now we come to a half dozen papers whose authors explore, from various perspectives, the unraveling of racial essentialism among American scholars and scientists in the twentieth century. We begin with three papers whose authors examine, from different perspectives, the importance of the work of the major architects of the so-called synthetic theory of evolution, which took shape in the period between the late 1930s and the early 1960s and undermined any support that the biological sciences, including genetics, might have had for racial essentialism.[11] The reasons were fairly complex, but they may be boiled down to one central notion: for the theorists of synthetic evolution, race was an open, dynamic category, a fluctuating population whose members shared certain genes in common.

Professor Vassiliki Betty Smocovitis gives us a sparking discussion of the problem of race in the botanical world—or, more precisely, the *nonproblem* of race in the botanical world. After discussing Darwin's attempts to deal with variation, heredity, and the species question in animals, Professor Smocovitis shows how complex these problems are in the world of plants. Plants are, after all, very different in many respects from animals: plants are generally developmentally simpler than animals, and are thus easier to hybridize than animals; plants crossbreed extensively, making reproductive isolation far more difficult than in the case of animals; plants have an indeterminate system of growth and development, which allows for a freer, more open-ended development of body parts and size; plants tend to be far more flexible phenotypically than animals—that is, more responsive to their environment; and plants sometimes multiply chromosome sets, and many species are capable of self-fertilization.

What all this adds up to is that however such categories as race or species as fixed entities help us understand animals, they put us at a loss in the plant world. In the plant world, categories such as race, species, and the like tend to confuse and befuddle more than clarify. One of the important comparisons Professor Smocovitis makes in her fine essay is the parallel work of the animal geneticist Theodosius Dobzhansky and his friend and colleague at Columbia University George Ledyard Stebbins, Jr., who worked on plant genetics. Both men contributed mightily to the development of the synthetic theory of evolution and thus to the undermining of racial essentialism. Both contributed to a unified synthetic theory of genetic evolution in the 1950s and 1960s. Each made major, seminal contributions to their respective fields of knowledge. In a very real sense we are left with Professor Smocovitis's very strong conclusion that in the end, race is a useful category only with respect to human civilization. Race, in short, is not and cannot be a category of modern biological science. In a word, it is a social construction.

Next comes the work of another key geneticist, Leslie Clarence Dunn, in a highly informative paper by Professor Melinda Gormley. For many years, Dunn was a productive mammalian geneticist at Columbia University, and thus a friend and colleague of both Dobzhansky and Stebbins. He was also involved in various liberal causes in the interwar and war years. For many years, he had argued in his political and social policy writings that cultural factors played a role in human evolution and variation—in other words, culture intervened with race, thus making racial essentialism, or scientific racism, deeply suspect if not downright wrongheaded. Dunn's opportunity for extensive work on human genetics in the experimental mode came with a long-delayed sabbatical in the mid-1950s that he took with his wife, Louise, and their son Stephen, an anthropologist. After some investigation, he settled on undertaking a genetic and cultural study of an isolated population of Jews in Rome, Italy—the longest surviving such aggregate of Jews outside of Palestine. Working with his son, Dunn was able to gather sufficient biological and cultural evidence that cultural factors did influence biological traits in specific populations. More specifically, they argued at the end of their work that they had demonstrated that the Jews of Rome's ghetto maintained their cultural and biological identity as a distinct subcommunity within the larger Italian and Catholic one of the eternal city and that social factors played a large role in this.

And how did the Dunns proceed? L. C. Dunn wished to study an isolated human community because he believed that isolation, geographical or cultural, provided the best clue as to how human races developed. Human races were produced by five distinct processes of evolution—mutation, selection, adaptation, migration, and isolation—and the last, isolation, was the most important, for it provided several factors that separated populations into races, including proximity, language, religion, education, and class or caste. At the same time, civilization, which was a relatively recent phenomenon in human history, tended to break down isolating barriers and promoted race mixing. It was usually cultural factors in recorded times that explained the segregation of populations, as in the examples of the forced cultural isolation of Americans of African and European ancestry or of caste isolation in India. And it was religion that most fascinated the Dunns as a potential isolating cultural factor in human populations. L. C. Dunn, the geneticist, deployed a technique that was then quite common in human studies. He would acquire blood samples from his population and test them for A, B, and O blood types, and then compare their characteristics to blood samples from other groups, perhaps with a similar genetic history or that lived nearby. This would yield access to human genotypes, thus permitting the evaluation of gene frequencies among the target population. Stephen Dunn, the anthropologist, interviewed members of the Rome ghetto and examined archival documents, especially records of births, marriages, and deaths, as well as deportations during the Nazi reign of terror. In both cases, father and son found it essential to use diplomacy, including food packages, to get their subjects to agree to participate in the study.

This Jewish community had been isolated from the surrounding Roman, Italian, and European populations for more than three centuries and then during a period of freedom from 1870 to1938, when the Fascist regime enacted laws that resegregated Jews from Christians. What anthropologist Stephen Dunn found was that intermarriage was quite common among the Roman Jews, whereas his geneticist father found that there were at least two significant gene frequency differences between the Jews and the larger Roman and Italian populations. Thus there were cultural and biological factors behind the isolation of the Jewish community and making them different from the Christian Italians who lived in Rome or in central Italy. L. C. Dunn thus argued that microevolution acted more or less the same on mice and human beings. Because humans were divided into races and subraces, and not into species and subspecies as were animals, cultural as well as biological factors played a role in isolation and the development of human races. Hence race was a flexible, dynamic category.

And what about the concept of race itself? Here Professor Farber offers us an acute analysis of changing notions of human race mixing, which ended up at approximately the same intellectual destination as the work of Stebbins and Dunn. Here the key architect was Dobzhansky, émigré from Ukraine who arrived to work in the famous Thomas Hunt Morgan's genetics laboratory in 1927. In accounting for the repudiation of scientific racism, or racial essentialism, in the postwar years, Professor Farber correctly notes that many historians have argued that the shift in scientific opinion on race primarily had its roots in social and ideological causes. Yet it is Professor Farber's contention that this shift in scientific opinion had strong scientific roots as well—that changes in strictly scientific notions about race removed the intellectual foundation and respectability of the scientific racism of the late nineteenth and early to mid-twentieth centuries.

Sentiment against the mixing of the races was extremely strong in late-nineteenth- and early twentieth-century America and among many American scientists. A clear example of this sort of sentiment was the eugenics movement, about which much ink has been spilled in the last several decades, beginning with Mark H. Haller's seminal 1963 publication, *Eugenics: Hereditarian Ideas in American Culture.* American eugenists insisted that when a superior race crossed with an inferior one, the result was always a biological disaster—a lowering of the standards of civilization. Here the assumptions were twofold: that there was a hierarchy of superior to inferior races, based on their accomplishments as "civilizations," and that the crossing of individuals from very different groups (e.g., Anglo-Saxon and African) would produce crosses that were asymmetrical or disharmonious. Many scientists in America and in Europe entertained these ideas quite seriously until the era of the Second World War, even though, as Professor Farber points out, the more careful reviews of the problem by qualified scientists such as Samuel J. Holmes of the University of California, Berkeley in 1936 conceded that the scientific evidence against race crossing was weak at best. An important scientific tribune of the arguments that race crossing did no harm was the Columbia University anthropologist Franz Boas. Boas subjected these ideas on the negative effects of race mixing to withering criticism from the standpoint of physical and cultural anthropology, fields in which he was equally a master. Boas insisted that so much race mixing had gone on in the past that no pure races existed. By the late 1920s and early 1930s, there was an entire "Boas school" of anthropologists, including Alfred L. Kroeber, Margaret Mead, Ruth Benedict, Alexander A. Goldenweiser, Robert H. Lowie, and Otto Klineberg, who subjected the theory of superior and inferior races—racial essentialism—to extensive criticism.[12]

In this essay, Professor Farber also notes the importance of the work of population geneticists on heredity, variation, and evolution in the 1930s and 1940s. In particular, it was the emergence of the Modern Synthesis, as biologists tended to dub it, that told the tale. The champions of the Modern Synthesis, along with those who plumped for natural selection in areas of biological science such as genetics, systematics, statistics, and field biology, among others, used the conception of race in accounting for the evolution of species. To these supporters, a race was a population or group of populations distinct from other populations of the same species. Breeding among the populations kept all races of the same species. But when isolation occurred, separation into distinctly different populations took place. In ways such as these, new races could emerge. Professor Farber points out that it was initially the work of Dobzhansky on races of fruit flies in the wild, first published in book form in 1937, that led the way to the reformulation of race as a dynamic, fluctuating concept allied with population genetics and the new randomized statistics of the 1930s. These changes had little or nothing to do with politics, the reaction against National Socialism, and the like. Hence, changes in notions of race had their definite roots in changes in scientific ideas in the 1930s and, to a lesser extent, in the 1940s. The political causes were more operational in the latter decade.

And what of the mental sciences—psychology and psychiatry—and race? It is perhaps not astonishing that our two chapters here are set on a national political stage. More than biology, psychology appears to the layperson as a science about which he or she is something of an authority and has an intelligible opinion. After all, those who argued for racial essentialism were especially concerned with the quality of mind of the various races. In my essay, I sketch a view of race and the intelligence quotient throughout the twentieth century as influenced by the changing tides of politics and science, and Professor Ben Keppel offers a more concentrated perspective on Robert Coles, a key psychiatrist who was especially active during what Professor Keppel calls the Second Reconstruction, stretching from the Supreme Court's decision in *Brown v. Board of Education* to the election of Ronald Reagan to the presidency in 1980.

There are interesting differences between the two stories told here. Generally speaking, the racial mental testers kept a distance between themselves and the objects of their investigations. Members of racial groups thus tested for "inferiority" of intellect were usually not persons with whom the testers came into close contact or got to know, at least apart from the testing situation. In that sense, testers were as "remote" from their subjects as the physical anthropologists of the previous century were from theirs in measuring, for example, average cranial capacity for a particular race. The testers made certain assumptions—that intelligence was inherited as a Mendelian or genetic trait, that it was fixed at birth, and that there was little in the examinee's milieu to change that ironclad destiny. Among the mental testers who assumed the superiority of the white race over peoples of color there was a large positivistic confidence in the metrics of intelligence testing and its various technical instruments: the intelligence quotient itself, the coefficient of correlation (often used to measure the similarity of ostensibly dissimilar things, as in physical and mental measurements), and the appropriateness of group measurements for determining individual intelligence quotients, among others.

On the other hand, psychiatrists such as Robert Coles and others with a similar approach, such as John Dollard, Allison Davis, David McClellan, John Whiting, Robert Sears, and many others of the social learning school of social psychology, had a closeness, or rather a virtual intimacy, with their subjects that the racial mental testers would have found strange, if not downright bizarre and unscientific. These psychiatrists interviewed their subjects, famously in Dollard's classic work, *Caste and Class in a Southern Town*, published in 1937. Coles followed suit when he began his work in the early 1960s. In his work as a participant-observer, Coles came to know his subjects—his interviewees—very well indeed. He and his wife lived in their communities in every sense of the word. He became intimately familiar with them as individuals. He could visualize particular persons, recall specific conversations, and understand, if not necessarily agree with, their views on matters large and small with regard to their lives, communities, and relations with individuals and groups of persons not like them at all. This sense of involvement between scientist and subject was quite rare for almost all the scientists and public advocates discussed in this essay, with one important exception: the dramatis personae from the Iowa Child Welfare Research Station who worked with the orphanage children came to see them as individuals over time, not as members of a group in a "group portrait," as other psychologists giving IQ tests so often did. That closeness of perspective likely mattered a good deal. Coming to understand one's subjects, one by one over a period of time, offered up rather different perspectives than lumping them together in groups of persons with seemingly similar traits and seeing them only at a distance, perhaps never coming into contact with them all, as was certainly the case with Arthur Jensen, who commented on others' studies, and Sir Cyril Burt, whose subjects in some cases may have well been fictive, especially in those studies in which Burt made the most hardcore hereditarian arguments.

The other point of comparison between my essay and Professors Keppel's is that what happened in the larger national culture and, above all, what took place in the areas of politics and group or racial relations mattered a good deal as an environment that helped shape, if not dictate in precise terms, what scientists had to say about the thorny problem of race.

And what are we to make of Professor Michael G. Kenny's fine paper on the genome and the reconfiguration of race? The new technologies associated with genomic research and explanation seem to play havoc with older categories of race and group identity. These ancient bromides and arguments fly out the window. If it is true that the new genomic science can divide, as Professor Kenny warns us, it is also true that it can unify and even perplex us. Human types that the eye cannot see can be derived from the new genomic science, and there is a bewildering infinity of individuals. Thus a "haplotype," the mixture of allelic states of a bundle of polymorphic markers on a chromosome or chromosomal region, may exist for each individual of a race or species so that no two individuals are exactly alike. There are, of course, markers in plentitude that stick together in groups and are thus inherited as a block—"haplogroups"—that constitute a population history, thus indicating the continuing relevance of population genetics and of the Modern Synthesis, as discussed by Professors Smocovitis, Gormley, and Farber, and as I implied in my essay. This is definitely an intimate view of one's subject and from the inside out, rather than from addressing the organism from the outside, so to speak. When we come to race, Professor Kenny warns us, it is enormously difficult to puzzle out whether the traditional racial categories make any sense at all, or whether the new genomic science and technology have made hash out of our time-honored categories of race. Furthermore, Professor Kenny continues, the traditional "races" are undergoing rather important revisions. All this makes racial essentialism highly questionable, as has the work in psychology and psychiatry, as Professor Keppel and I discuss, not to mention the work in genetics and evolution chronicled by Professors Smocovitis, Gormley, and Farber.

We may or may not have come to a point in our national history in which racial essentialism as a cultural, political, and social tradition—or bromide—has become problematic, questionable, or even eliminated. But it would appear that the scientific justification for regarding the different races of our species as arranged in a hierarchy of superior to inferior races has, in a word, evaporated. This is not to say that no one will try to revive scientific theories of racial superiority and inferiority again. Anything along these lines is possible, and the history of scientific racism shows over and over again the propensity to believe in that argument, regardless of what sort of information one can adduce on its behalf. But we would conclude by saying that the case has become very difficult to make on scientific grounds, for the very grounds of "racial science" have shifted, and dramatically so, especially in the several decades since the 1930s.

Notes

1. What had existed before in European and North American society was the notion of a society of orders, which dated from the Middle Ages and was a very different kind of idea of the arrangement of peoples.
2. Juergen Habermas, *The Structural Transformation of the Public Sphere: An Inquiry into a Category of Bourgeois Society* (Cambridge, MA: MIT Press, 1989). See also Gordon Wood, *The Radicalism of the American Revolution* (New York: Alfred A. Knopf, 1992); and Paul Starr, *The Creation of the Media* (New York: Free Press, 2004).
3. George H. Daniels, *American Science in the Age of Jackson* (New York: Columbia University Press, 1968) analyzes the decline of "Baconian" science by the mid-1840s, which signaled the end of the doctrine of design and the emergence of scientific materialism, including the idea of a statistical norm.
4. Peter Gay, *The Enlightenment: An Interpretation*, vol. 1, *The Rise of Paganism* (New York: Alfred A. Knopf, 1966), passim; and Robert R. Palmer, *The Age of the Democratic Revolution: A Political History of Europe and America, 1760–1800* (Princeton, NJ: Princeton University Press, 1959–1964), 1:passim.
5. Among the many first-rate books that discuss scientific theories of race in nineteenth-century America are William B. Stanton, *The Leopard's Spots: Scientific Attitudes Towards Race in America, 1815–1859* (Chicago: University of Chicago Press, 1960); George M. Fredrickson, *The Black Image in the White Mind: The Debate on Afro-American Character and Destiny, 1817–1914* (New York: Harper and Row, 1971); Thomas F. Gossett, *Race: The History of an Idea in America* (Dallas: Southern Methodist University Press, 1963); and the many books by George W. Stocking, Jr., beginning with *Race, Culture, and Evolution: Essays in the History of Anthropology* (New York: Free Press, 1968).
6. Frank Tannenbaum, *Slave and Citizen: The Negro in the Americas* (New York: Alfred A. Knopf, 1947) is the seminal work on this issue. See also Carl N. Degler, *Neither Black nor White: Slavery and Race Relations in Brazil and the United States* (New York: Macmillan, 1971).
7. In general, see Gary B. Nash, *Red, White, and Black: The Peoples of Early North America* (Upper Saddle River, NJ: Prentice-Hall, 2002).
8. Among the many books that could be cited on these matters are Winthrop Jordan, *Black Over White: American Attitudes toward the Negro, 1550–1812* (Chapel Hill: University of North Carolina Press, 1968); David B. Davis, *The Problem of Slavery in the Age of Revolution, 1770–1823* (Ithaca, NY: Cornell University Press, 1975); Davis, *The Problem of Slavery in Western Culture* (Ithaca, NY: Cornell University Press, 1986); Ira Berlin, *Many Thousands Gone: The First Two Centuries of Slavery* (Cambridge, MA: Harvard University Press, 1998); and Berlin, *Generations of Captivity: A History of African-American Slaves* (Cambridge, MA: Harvard University Press, 2003).
9. Among works that could be consulted are Francis Paul Prucha, *American Indian Policy in the Formative Years: The Indian Trade and Intercourse Acts, 1780–1834* (Cambridge, MA: Harvard University Press, 1962); Prucha, *The Great Father: The United States Government and the American Indians* (Lincoln: University of Nebraska Press, 1986); and Prucha, *A Bibliographical Guide to the History of Indian-White Relations in the United States* (Chicago: University of Chicago Press, 1977).

10. Edward J. Larson, *A Magnificent Catastrophe: The Tumultuous Election of 1800, America's First Presidential Campaign* (New York: Free Press, 2007).
11. See, for example, Hamilton Cravens, *The Triumph of Evolution: American Scientists and the Heredity-Environment Controversy, 1900–1941* (Philadelphia: University of Pennsylvania Press, 1978), 156–210; and Vassiliki Betty Smocovitis, *Unifying Biology: The Evolutionary Synthesis and Evolutionary Biology* (Princeton, NJ: Princeton University Press, 1996), passim. An important contemporary statement of the synthetic theory of evolution is Theodosius Dobzhansky, *Mankind Evolving* (New Haven, CT: Yale University Press, 1960).
12. See, for example, Cravens, *Triumph of Evolution*, 107–210 (see n. 11); and Stocking, *Race, Culture, and Evolution*, 270–307 (see n. 5).

Slavery in the Election of 1800

Edward J. Larson

In 1800, the aristocratic slave state of Virginia stood at the Republican epicenter of the first truly partisan political campaign for president. Republican members of Congress had nominated Virginia plantation owner Thomas Jefferson, the author of the Declaration of Independence, which had proudly declared that "all men are created equal," as their party's candidate for president. Although his partisans then called themselves Republicans, they were often denounced as Democrats by their conservative Federalist opponents and would eventually adopt that name for themselves.

As he had in the 1796 presidential election after George Washington declined to stand for a third term, Jefferson faced John Adams of Massachusetts, his former friend and fellow Revolutionary Era patriot. That first contested election had not involved a full-blown partisan campaign, but the tight finish pushed each side to organize itself into a formal national party for the next election. Adams had narrowly won in 1796 and was renominated four years later by the Federalist caucus in Congress. Both men had held office during the Washington administration, with Jefferson serving as the secretary of state and Adams as vice president. Events had driven them apart, however. By 1800 they were bitter partisan foes.

Jefferson's Second Revolution

The differences dividing Adams and Jefferson reflected a deepening ideological rift that divided Americans. Even as the Washington administration first took shape, the people and their leaders vigorously debated various issues regarding the authority of the national government and the balance of power among its branches and between it and the states. Whether the national government could charter a bank and thus create a national banking system became an especially heated topic, for example. Adams and those calling themselves Federalists saw a strong central government led by a powerful president as vital for a prosperous, secure nation. By preempting state regulation of business, they hoped that the central government could provide a basis for a truly national market economy that would facilitate the development of manufacturing and commerce.

Strong proponents of this nationalistic viewpoint, like Alexander Hamilton of New York, who favored transferring virtually all power to the national government and consolidating it in a strong executive and aristocratic senate, become known as the High Federalists. At the Constitutional Convention in 1787, Hamilton had unabashedly depicted the monarchical British government as "the best in the world" and famously proposed life tenure for the president and senators.[1] Befitting their support for business, Hamilton and the High Federalists favored close trading ties with Britain, which remained the nation's principal trading partner even after the Revolution. The new government's resulting pro-British foreign policy inevitably resulted in tensions between the United States and France, especially after the French Revolution resulted in open war between Europe's ancient monarchies, led by Britain, against Europe's rising republican regimes, led by France.

Jefferson and his emerging Republican faction viewed such thinking as inimical to individual freedom. A devotee of enlightenment science, which emphasized reason and natural law over revelation and authoritarian regimes, Jefferson trusted popular rule and distrusted elite institutions, particularly entrenched aristocracies and organized religions that seemed to subvert the individual. "The will of the majority, the natural law of every society, is the only sure guardian of the rights of men," Jefferson wrote in 1790.[2] He had no love of Britain or its institutions and great hope for the liberating effect of the French Revolution. Indeed, he viewed the upheaval in France as a European manifestation of the American revolutionary spirit.

In contrast, Adams and the Federalists tended to distrust the common people and instead placed their faith in the empowerment of what they saw as a natural aristocracy, though one that should be restrained by civil institutions such as those provided by a written constitution with checks and balances. "The voice of the people has been said to be the voice of God, and however generally this maxim has been quoted and believed, it is not true," Hamilton reportedly told the Constitutional Convention. "The people are turbulent and changing; they seldom judge or determine right."[3]

Though more moderate in his views than Hamilton, Adams agreed that every nation needed a single strong leader who could rise above and control self-interested factions. Neither an aristocracy nor a democratic majority would safeguard individual rights, he believed. Only a disinterested chief executive—the fabled philosopher king of old—would protect liberty and justice for all. Accordingly, Adams supported a strong presidency. Although preferring an elected leader to a hereditary one, Adams's thinking leaned too much toward monarchism for Jefferson to stomach, especially when Federalists in high positions around Adams openly praised the balanced British constitution with its hereditary House of Lords, limited House of Commons, and still-powerful king. As Washington's treasury secretary,

Hamilton pushed a centralizing, pro-business program of internal taxes, protective tariffs, a national bank, and close trading ties with Britain. He viewed them as essential for national power, prestige, and prosperity.

From within the administration, Secretary of State Jefferson opposed all these policies as corrosive of individual liberty and equal opportunity. Even more, he feared that they would undermine popular rule by creating an aristocracy of wealth in America—a homegrown elite. He did not want the United States simply to become a better Britain, with its concentrated wealth and power. He dreamed of something new under the sun in America—a land of free, prosperous farmers and workers. By 1792 Virginia congressman James Madison, who acted on Jefferson's behalf in such matters, was calling for a formal political party to oppose Hamilton and his followers. Organization on the left precipitated organizing on the right, and soon America's characteristic two-party system emerged in the rise of partisan newspapers, the coordination of party voting by members of Congress, party endorsements for political candidates, and state party structures.

In 1800 Jefferson and the Republicans ran on a platform of states' rights and individual liberty. They accused the Federalists of trying to restore a monarchy in America, with Adams or Hamilton as the king. During Adams's first term in office, a perceived threat from revolutionary France to American interest had led to naval clashes between the United States and France on the high seas, a military buildup at home, the imposition of heavy war taxes, and the enactment of the Alien and Sedition Acts to address concerns about domestic security in wartime. Republicans pointed to these measures as evidence of the Federalists' monarchical designs. Claiming that freedom was at stake, Jefferson likened his 1800 presidential campaign to the patriot cause of 1776 and hailed his victory as "a revolution in the principles of our government."[4]

Race and slavery represented something of an eddy within the major currents dividing the two parties. There were Federalists in all parts of the country, of course, but the party's heart lay in the North with the manufacturing and commercial elites. Their businesses relied on free workers and small farmers. Northern evangelicals leaned heavily toward the Federalist side as well, and these evangelicals were turning against slavery as a violation of divine law. Adams had been the only signer of the Declaration of Independence never to own a slave, and he remained personally opposed to the institution. At the Constitution Convention, Gouverneur Morris, who served as a prominent Federalist senator from New York in 1800, spoke most strongly against slavery, calling it "a nefarious institution" and "the curse of heaven on the states where it prevailed."[5]

In the Republican South, in contrast, agriculture typically was conducted on large plantations, with African slaves supplying much of the labor. Despite their eloquent pleas for liberty and equal opportunity, Jefferson, Madison, and other Southern Republican leaders owned large plantations and numerous slaves. Jefferson's rhetoric often appealed to free workers and new immigrants in the North, but the base of his support remained in the slave states of the South, where only free, property-owning men could vote. In 1796 Adams and the Federalists had carried every free state north of the Mason-Dixon Line except Pennsylvania; Jefferson and the Republicans had carried all the slave states south of Maryland. When Southern Jeffersonians spoke warmly of liberty and equal opportunity, they clearly meant it for white men. Their words inevitably reached other ears, however.

Revolution within a Revolution

Gabriel was a large man with big plans. Like so many other Virginians in 1800, he dreamed of freedom. Born in 1776, Gabriel became a blacksmith and apparently worked at various sites in and around the state capital of Richmond. He stood well over six feet tall, had a powerful build, and commanded respect from his peers. Those peers, however, were African American slaves, and Gabriel wanted more for them—and for himself—than bondage. As a semi-itinerant craftsman, he inevitably associated with free blacks and white artisans. Most of his earnings, however, went to his owner, Thomas Prosser, who engaged in the then-common practice of letting some of his skilled slaves work independently. The taste of personal liberty Gabriel enjoyed may have increased his hunger for complete emancipation; it certainly allowed him to experience the election-year tumult in Virginia and to hear radical Republican calls for liberty.

One contemporary newspaper account described Gabriel as "a fellow of courage and intellect above his rank in life."[6] He could read and write. He also had a temper. In 1799 Gabriel was caught stealing a pig from a white overseer named Absalom Johnson. Gabriel bit off the Johnson's ear in the ensuing scuffle and was branded on the left hand for his crime. At some point shortly after this incident, Gabriel decided that he had had enough of slavery and scraping for food. He began plotting a massive insurrection designed to win freedom for himself and other slaves in southeastern Virginia. Events in revolutionary France and the wealthy French Caribbean colony of St. Domingue on Hispaniola Island, which slaves had taken over in 1793 and proclaimed their freedom, may have inspired him.

During the spring and summer of 1800, Gabriel conspired with other slaves in the region—probably too many—and became their general. Others served as captains and sergeants. Two radical white French immigrants and some free African Americans allegedly also participated in the plotting. Gabriel forged swords from sickles for his men and made five hundred bullets for their few guns. He claimed that thousands of slaves would rise on his call and hoped that free blacks and poor whites would rally to his banner. Inverting Patrick Henry's famous cry, Gabriel crafted a flag that read, "Death or Liberty."[7] According to courtroom testimony from participants, Gabriel at times demanded death for all the whites in the area except "Quakers, Methodist, and French people"[8]—three groups widely perceived to oppose slavery. At other times he spoke of dining and drinking "with the merchants of the city" once they agreed to end slavery.[9] In either event, Gabriel anticipated a bloody fight for freedom. "It is unquestionably the most serious and formidable conspiracy we have ever known of this kind," Virginia's Republican governor, James Monroe, reported to his close friend and political ally Jefferson in September, after the plot became public.[10]

The uprising might have succeeded up to a point had nature not intervened. The plan was as follows: On the evening of Saturday, August 30, the slaves around Prosser's plantation, six miles north of Richmond, would rise up, kill Gabriel's chief tormentors, Prosser and Johnson, and secure guns from a nearby tavern. Other slaves would join them for a midnight march on Richmond. Once in the city, some participants would set the warehouse district on fire to divert attention while others captured the state capitol, armory, treasury, and governor's mansion. They would use the state's guns as their weapons and distribute the state's treasure among their troops. Gabriel planned to take the governor hostage and bargain for emancipation. By some accounts, Gabriel harbored hopes that Monroe, who, as an American ambassador in Paris, once embraced the revolutionary regime in France, might prove cooperative. From this point on, Gabriel's plans are too vague to reconstruct, but they went awry from the outset.

"Upon that very evening" of August 30, Republican printer John Thomson Callender reported in a letter to Jefferson, "there came on the most terrible thunder storm, accompanied with an enormous rain, that I ever witnessed in this state." The road between Prosser's plantation and Richmond became impassible. Only a few local slaves turned out. "They were deprived of the juncture and assistance of their good friends in this city, who could not go out to join them," Callender noted.[11] Gabriel postponed the uprising for a day and sent his followers home.

Crushing the Revolution

Rumors of a slave conspiracy had circulated in Virginia throughout the summer, but Monroe had dismissed and suppressed them. In 1800 Republican leaders from Jefferson on down sought to discourage and deny threats of revolutionary activity at home so as not to feed Federalist fearmongering on the issue of domestic security. They spoke and acted like moderate Federalists and appealed to their more radical supporters only for votes in contested elections. Threats of a slave revolt posed particular problems for Republicans. Southern Federalists had long warned that Republican calls for liberty and equality could stir up the slaves. In 1799, following the XYZ Affair, South Carolina Federalist congressman Robert Goodloe Harper published an open letter to his constituents warning that France "was preparing to invade the southern states from St. Domingo with an army of [freed] blacks, which was to be landed with a large supply of officers, arms, and ammunition, to excite an insurrection among the Negroes."[12]

After the aborted rising on August 30, when it became clear that an insurrection might happen, Monroe responded swiftly as if to reassure citizens that they could trust Republicans to maintain order. With critical state elections scheduled for October in the slave states of Maryland and South Carolina, Monroe would not run the risk of a violent insurrection by blacks in Republican-ruled Virginia. "The scenes which are acted in St. Domingo," Monroe later wrote to his state military commander, "must produce an effect on all people of color in this and the states south of us, more especially our slaves, and it is our duty to be on guard to prevent any mischief resulting from it."[13]

After Gabriel dismissed those few conspirators who managed to assemble at the appointed time despite the rain, one of them—perhaps unnerved by the delay—told his owner about the planned insurrection and its leader's name. That owner raised the alarm. Other owners soon heard about it from their slaves as well. Within hours, members of the county militia began hunting down the conspirators, and Gabriel fled for the coast.

Word of these developments reached the governor in Richmond on August 31. He promptly moved the public arms to a secure location. As the full extent of the conspiracy became clear, Monroe called out various local units of the state militia, and Richmond took on a military face. Soldiers began systematically rounding up slaves and arresting those thought to have joined the conspiracy. "There has been great alarm here of late at the prospect of an insurrection of the Negroes in this city and its neighborhood," the governor wrote to Jefferson on September 9. "About thirty are in prison who are to be tried on Thursday, and others are daily discovered and apprehended in the vicinity of the city. . . . It is the opinion of the magistrates who examined those committed that the whole, very few accepted, will be condemned."[14] Monroe promised a $300 reward for anyone who captured Gabriel, plus a full pardon if an accomplice did the deed.

Under Virginia law, accused slaves appeared before a special court composed of five judges, all of whom had to agree on any punishment. Except for the governor's power to pardon, the court's decision was final. In cases of conspiracy, rebellion, or insurrection, the court could impose the death penalty and order an immediate public hanging. On the initial day of court proceedings, which occurred less than two weeks after the first arrests, the court sentenced all six defendants to die at dawn or noon the next day. Nine more men were hung the following week. At these executions, troops surrounded the scaffold to keep the hostile crowd from assaulting the convicted slaves before the executioners could kill them. "The whole state has been in consternation," a High Federalist commented smugly. "Courts are sitting, trials are taking place, and the gallows are in full operation."[15]

The Politics of Revolution and Reprisals

By mid-September Monroe began to question the wisdom of further reprisals and perhaps fear a political backlash from too many hangings. "While it was possible to keep [the threat of an insurrection] secret, which it was till we saw the extent of it, we did so," he explained in a letter to Jefferson on September 15. "But when it became indispensably necessary to resort to strong measures with a view to protect the town, the public arms, the treasury, and the jail, which were all threatened, the opposite course was in part taken. We then made a display of our force." By the time Monroe wrote these words, the state had already executed ten conspirators. He predicted that up to forty more would hang unless he intervened with pardons. "When to arrest the hand of the executioner is a question of great importance," Monroe wrote to Jefferson. "I shall be happy to have your opinion on these points."[16]

For Jefferson and Monroe, how to handle the conspirators was more than simply a moral or ethical question. From a political standpoint, Virginia Republicans needed to display sufficient toughness to assure frightened citizens (particularly in key Southern states) that a Jefferson administration would keep the peace and suppress leveling insurrections. At the same time, however, showing excessive harshness could alienate voters opposed to slavery or those sympathetic to slaves caught up in the aborted insurrection. Pennsylvania posed a particular problem in this respect. A hotbed of radical

Republicanism and a center for America's growing abolition movement, it would hold its state legislative election in four weeks, with fifteen electoral votes hanging in the balance. Rumors had already placed one of Gabriel's alleged French coconspirators in Philadelphia.

By mid-September the Federalist press began capitalizing on the episode by presenting it as a natural consequence of Republicanism run riot. "The sound of French Liberty and Equality in the ears of these Blacks led them to this desperate measure," the author of one widely reprinted article charged.[17] "Behold America, the French doctrine of Insurrection!" another writer exclaimed.[18] Some articles criticized Republican printers for filling the heads of slaves with notions of freedom. "The slave holders in our county no longer permit the *Aurora* and other *Jacobin* papers to come into their homes as they are convinced the late insurrection is to be attributable entirely to this source," a correspondent supposedly from Virginia wrote in a letter appearing in Federalist papers across the country.[19]

Many articles blamed Jefferson's egalitarian rhetoric for the slaves' actions. "Truly Mr. J.," read a typical charge, "should the business end in massacre, you and your disciples are the men who are the cause of it, and for every outrage and murder the Negroes may commit, you stand accountable."[20] After leveling similar accusations against Virginia Republicans in general, a Connecticut newspaper urged state voters "to tread down Jacobin philosophy and fractious reformation; to support by constant precept and example the dominion of religion, order, and law; and to cling solely to the Federal[ist] government as the only rock of their stability."[21] Radical Federalist printer William Cobbett commented, "The late revolt . . . amongst the Negroes of Virginia . . . will make Jefferson and his party very cautious how they do any act which may stir the sleeping embers of that alarming fire which, were it once rekindled, would probably make all the southern states what Hispaniola now is."[22]

The Federalist press also indicted slaveholding Virginia Republicans for hypocrisy in their handling of the affair and suggested that the conspirators died for Jefferson's sins. "He who effects to be a Democrat and is at the same time an owner of slaves, is a devil incarnate," declared one Federalist writer. "Democracy therefore in Virginia is like virtue in hell."[23] One Federalist newspaper depicted the conspiracy as "shallow" and easily suppressed,[24] which carried the implication that Virginia Republicans overreacted in their response to it. Several articles reminded readers of Jefferson's earlier praise for "the boisterous sea of liberty," and, in light of the threatened slave revolt, contrasted that image with the Federalist promise of ordered freedom.[25] "If anything will correct and bring to repentance old hardened sinners in

Jacobinism, it must be an *insurrection of their slaves*," a writer at the Federalist *Boston Gazette* observed. "One old experienced statesmen like John Adams, who honestly tells man how wicked they are and that nothing will keep them in good order but the powerful restraints of a strong government, is worth all the speculative philosophers from Thomas Jefferson down."[26]

Potentially the most explosive evidence to emerge from the conspiracy trials in Virginia involved the testimony of multiple witnesses that two white Frenchmen helped Gabriel. In their confessions, some of the conspirators named at least one of these alleged collaborators and suggested that both played major roles in the effort. Federalist newspapers latched on to this testimony and published it along with accusations that the revolutionary rhetoric of domestic Republicans inspired the slaves to revolt. Anything linking French Jacobins to domestic instability helped justify the Federalists' Alien Act and counter Republican criticisms of it.

Republicans tried to deflect these charges by denying them. The nation's flagship Republican newspaper, the *Aurora*, dismissed published reports that radical Republicans instigated the Virginia slave conspiracy as "wholly false" and suggested that Federalist policies such as conducting trade talks with Toussaint Louverture, the black ruler of St. Domingue, contributed more to the unrest than anything Republicans said or did.[27] "While our administration was encouraging . . . revolt and trading with Toussaint in the West Indies, what could be expected from the unfortunate Blacks and slaves in our states from the example?" a writer at the newspaper asked.[28]

Not withstanding undisputed trial testimony to the contrary, Republicans maintained that Gabriel and his fellow slaves acted without outside assistance. According to the *Aurora*, "There was not so much as the slightest foundation for suspecting any Republican American or any Frenchman" played a role in the affair.[29] Republican newspapers around the country reprinted this denial and made it their own. When pressed on the issue shortly before the fall election, Monroe asserted, "According to our present information, the conspiracy was quite a domestic one, conceived and carried to the stage at which it was discovered by some bold adventurers among the slaves." As if for emphasis, he added, "If white men were engaged in it, it is a fact of which we have no proof."[30] Based on his study of the episode, however, historian Douglas Egerton concluded that Monroe probably had received at least some evidence of white participation but suppressed or destroyed it. Virginia officials neither pursued the accusations of the Frenchmen's involvement nor indicted any white people in the case.

By all accounts, Federalists believed what their papers said about Republican complicity in the conspiracy. They made similar comments in private. "I doubt not that the eternal clamor about liberty in Virginia . . . has matured the event which has happened," a leading Federalist diplomat wrote about the affair in a letter to the president's son John Quincy Adams.[31] "In Virginia, they are beginning to feel the happy effects of liberty and equality," another prominent Federalist added. "The reports from that quarter say it was planned by Frenchmen, and that all the whites, save the French, were to have been sacrificed."[32]

American Sphinx

By the early fall of 1800, Jefferson and Monroe viewed virtually everything—even a desperate slave conspiracy and its violent suppression—in political terms. On September 20, Jefferson gave a cautious reply to Monroe's question about when to stop the hangings. Jefferson always hedged on the issue of slavery, so much so that one of his best biographers, Joseph Ellis, called him the American sphinx. Jefferson owned slaves throughout his adult life and treated them like property. In his draft for the Declaration of Independence, however, Jefferson listed the institution of slavery as one of the usurpations by the British monarch that justified the American Revolution. A decade later, in *Notes on Virginia*, Jefferson wrote of his hope, "under the auspices of heaven, for a total emancipation [of the slaves] with the consent of the masters."[33] In that same book, however, Jefferson suggested that Africans and Europeans constituted separate species that did not spring from common ancestors. In line with a growing body of Southern racist science, he depicted the two races as separate creations. By 1800, however, whispered rumors circulated that the widowed master of Monticello had sexual relations with his female slaves. Slavery, and the treatment of convicted slave conspirators during a political season, presented Jefferson with hard issues.

"Where to stay the hand of the executioner is an important question," Jefferson wrote to Monroe. Virginians would differ in their answer, Jefferson observed, but the political ramifications reached beyond Virginia. "The other states and the world at large will forever condemn us if we indulge in a principle of revenge or go one step beyond absolute necessity," he wrote. "They cannot lose sight of the rights of the two parties and the object of the unsuccessful one. Our situation is indeed a difficult one."[34]

Jefferson suggested exporting the convicted slaves out of the country rather than hanging them. "I hazard these thoughts for your consideration only," the cautious candidate added, "as I should be unwilling to be quoted in the case."[35] After Jefferson's letter reached Monroe, most convicted conspirators received outright pardons on the court's recommendation of mercy or were "reprieved for transportation"[36] to Spanish Louisiana. The political storm passed with minimal impact on the election.

Before the end of the affair, however, Virginians with power demanded at least one more execution. The $300 bounty offered for Gabriel's capture had worked. On September 23 in Norfolk, a slave with no part in the conspiracy betrayed Gabriel's hiding place to the local sheriff, who arrested him and sent him to the state capital for trial. Gabriel reached Richmond in irons on September 27. His brief public trial was held nine days later. Gabriel was the sixth man tried that day and the only one sentenced to die without a recommendation of mercy. The trial drew a large crowd.

Gabriel sat silently as three of his former followers placed him at the center of the conspiracy. They gained their lives for their testimony while Gabriel lost his. Gabriel only spoke after the court sentenced him to die the next day. He asked for a delay of three days so that he could hang with six of his coconspirators, who were previously scheduled for execution on October 10. The court granted his last request. In all, Virginia hung twenty-six slaves for their role in the conspiracy, with Gabriel being the last of them to die in Richmond.

Slavery and Politics in 1800

Despite their partisan wrangling over the causes and handling of the Virginia slave conspiracy during the campaign of 1800, neither Federalists nor Republicans spoke substantively to the underlying issue of slavery. Even though most Northern states had abolished slavery by 1800, it remained deeply entrenched in the South. Neither party could hope to win the presidency if it took a strong stand on slavery, so they both equivocated on what had already emerged as the most divisive topic in American politics.

Both parties were deeply split by the issue. Slavery disgusted Adams; he once called it "an evil of colossal magnitude."[37] Yet he included three slave owners in his five-member cabinet and his hope for reelection rode on winning electoral votes from three slave states: Maryland, Delaware, and South Carolina. Many northern High Federalists opposed slavery on moral or religious grounds, yet their faction's favored candidate for president, Charles Coteworth Pinckney, possessed vast slave plantations and, as a delegate to the Constitutional Convention, led the successful effort to ensure that the Constitution protected the right of states to maintain slavery. If the Constitution "should fail to insure some security to the southern states against an emancipation of slaves," Pinckney told his fellow delegates, he "would be bound by this duty to his state to vote against [it]."[38] After initially demanding that slaves be counted like free people for purposes on computing the number of a state's representatives in Congress and electoral votes for president—all the while acknowledging that slaves themselves could not vote—Pinckney and his allies at the Constitutional Convention eventually settled for counting each slave as three-fifths of a free person for such purposes.

The Republican Party encompassed a similar diversity of views on slavery, from the ardent support for it expressed by many party leaders in the Deep South through Jefferson's tortured acquiescence of the practice to the fevered abolitionism of prominent Northern Republicans such as Pennsylvania congressman Albert Gallatin. "Slavery is inconsistent with every principle of humanity, justice, and right," Gallatin had written in a 1793 legislative report,[39] yet he served as Jefferson's point man in Congress during the 1800 election.

In 1800 none of the national candidates questioned the right of states to authorize slavery or proposed that the government do anything to discourage slavery or restrict the slave trade. At most, partisan pundits postured on the edges of these explosive issues. In articles and pamphlets addressed to voters in their region, for example, Northern Republicans frequently reminded voters of Jefferson's expressed hope for gradual emancipation. "The spirit of the master is abating, that of the slave rising from the dust," Jefferson had written in "Notes on the State of Virginia"[40] as part of a passage critical of slavery that Adams once described as "worth diamonds."[41] Southern Federalists used the same passage, which warned of dire consequences if slavery did not end, to turn white voters in the South against Jefferson. In a published campaign address, a Rhode Island Federalist ridiculed the calls for liberty and equality coming from Virginia Republicans. Should New Englanders "take lessons upon those subjects from the state of Virginia . . . where slavery constitutes a part of the policy of the government?" he asked.[42]

A late-September exchange over Gabriel's Revolt between Philadelphia's two leading partisan newspapers showed just how far rhetoric had departed from reality. "The insurrection of the Negroes in the southern states, *which appears to be organized on the true French plan*, must be decisive with every reflecting man in those states of the election of Mr. Adams and Gen. Pinckney," the writer of an essay in the Federalist *Gazette of the United States* asserted.[43] Scared for their safety, white Southern voters would now turn to Federalists for security, the essayist suggested. "We augur better things from this unhappy but, thank God, partial revolt," a writer at the Republican *Aurora* replied a day later. "We augur from it . . . the election of [Thomas Jefferson] whose whole life has been marked by measures calculated to procure the emancipation of the Blacks."[44] Whether true or not, that was what Pennsylvania Republicans wanted to hear about their candidate.

Although such dueling comments probably did not change many minds, they likely spoke to each party's local political base.

By 1800 slavery had already become a potent political issue. Each party sought to use both sides of this divisive issue for short-term political gain without addressing the underlying matter of slavery in any substantive way. The pattern of playing politics with race had already emerged as a feature of American campaigns. As would happen in future elections, race and slavery played a decisive role in the outcome. The ultimate contest again proved so close that, without the extra electors given to slave states by the Three-fifths Compromise, Adams would have won. As it turned out, because of the added electoral votes held by the slave states in Jefferson's Southern base, the Virginian gained a narrow victory. Southern Democrats would hold power in the national government for most of the ensuing sixty years, entrenching slavery in the South even as moral and political objections to the practice grew in the North and West. Jefferson would later refer to the sectional political split over slavery, when it manifested itself in the 1820 Missouri Compromise, as "a fire bell in the night."[45] That damning signal, which would rip the country apart in 1860 presidential election, was already ringing softly in the campaign of 1800.

Notes

1. Edward J. Larson and Michael P. Winship, *The Constitutional Convention: A Narrative History from the Notes of James Madison* (New York: Modern Library, 2005), 49.
2. Thomas Jefferson to citizens of Albemarle County, Virginia, 12 February 1790, in *The Papers of Thomas Jefferson*, ed. Julian P. Boyd et al. (Princeton, NJ: Princeton University Press, 2005), 16:179.
3. Larson and Winship, *Constitutional Convention*, 50 (see n. 1).
4. Thomas Jefferson to Spencer Roane, 6 September 1819, in *The Works of Thomas Jefferson*, ed. Paul Leicester Ford (New York: G. P. Putnam's Sons, 1905), 12:136.
5. Larson and Winship, *Constitutional Convention*, 112 (see n. 1).
6. "A Fellow of Courage," *New York American Citizen*, September 25, 1800, 2.
7. "The Trial of Gabriel," *Journal of the Senate of Virginia* (Richmond: Nelson, 1800), 32.
8. Ibid.
9. Ibid.
10. James Monroe to Thomas Jefferson, 15 September 1800, in *The Writings of James Monroe*, ed. Stanislaus Murray Hamilton (New York: G. P. Putnam's Sons, 1900), 3:208.
11. James Thomson Callender to Thomas Jefferson, 13 September 1800, in *Papers of Thomas Jefferson*, 32:136 (see n. 2).
12. Robert G. Harper, "Letter to Constituents," *Philadelphia Gazette of the United States*, April 3, 1799, 3.
13. James Monroe to General Mathews, 17 March 1802, in Douglas R. Egerton, *Gabriel's Rebellion: The Virginia Slave Conspiracies of 1800 and 1802* (Chapel Hill: University of North Carolina Press, 1993), 47.
14. James Monroe to Thomas Jefferson, 9 September 1800, in *Writings of James Monroe*, 3:205 (see n. 10).
15. Robert Troup to Rufus King, 1 October 1800, in *The Life and Correspondence of Rufus King*, ed. Charles R. King (New York: G. P. Putnam's Sons, 1896), 3:316.
16. James Monroe to Thomas Jefferson, 15 September 1800, in *Papers of Thomas Jefferson*, 32:144–45 (see n. 2).
17. *Portsmouth, New Hampshire Gazette*, September 30, 1800, 3.
18. *Philadelphia Gazette*, October 24, 1800, 3.
19. *Philadelphia Gazette of the United States*, October 1, 1800, 3.
20. *Philadelphia Gazette of the United States*, September 18, 1800, 3.
21. *Norwich (CT) Courier*, October 15, 1800, 2.
22. William Cobbett, *Porcupine's Works* (London: Cobbett & Morgan, 1801), 12:141.
23. *New London, Connecticut Gazette*, October 8, 1800, 1.
24. *Windham (CT) Herald*, October 2, 1800, 2 (reprinted from *Richmond, Virginia Gazette*, September 12, 1800).
25. Thomas Jefferson to Philip Mazzei, 24 April 1796, in Thomas Jefferson, *Writings* (New York: Library of America, 1984), 1037. See, for example, *New York Commercial Advertiser*, October 1, 1800, 3.
26. *Boston Gazette*, October 9, 1800, 3.
27. *Washington (PA) Herald of Liberty*, October 27, 1800, 3 (reprinted from *Philadelphia Aurora*, October 7, 1800).

28. *Alexandria (VA) Times and District of Columbia Daily Advertiser*, October 14, 1800, 2 (reprinted from *Philadelphia Aurora*, date unknown).
29. See, for example, *Hartford (CT) American Mercury*, October 16, 1800, 3 (reprinted from *Philadelphia Aurora*, date unknown).
30. James Monroe to John Drayton, 21 October 1800, in *Writings of James Monroe*, 3:217 (see n. 10).
31. William Vans Murray to John Quincy Adams, 9 December 1800, "Letters of William Vans Murray to John Quincy Adams, 1797–1803," *Annual Report of the American Historical Association* (1912): 663.
32. Robert Troup to Rufus King (see n. 15).
33. Thomas Jefferson, "Notes on the State of Virginia," in *Writings* (New York: Library of America, 1984), query 28, 289.
34. Thomas Jefferson to James Monroe, 20 September 1800, in *Papers of Thomas Jefferson*, 32:160 (see n. 2).
35. Ibid.
36. Egerton, *Gabriel's Rebellion*, 112 (see n. 13). This term is from the time for the export of a convicted slave.
37. John Adams to William Turner, 20 November 1819, Adams Family Papers, microfilm reel 124, Massachusetts Historical Society, Boston.
38. Larson and Winship, *Constitutional Convention*, 106 (see n. 1).
39. Henry Adams, *The Life of Albert Gallatin* (Philadelphia: Lippicott, 1880), 86.
40. Jefferson, "Notes on the State of Virginia" (see n. 33).
41. John Adams to Thomas Jefferson, 22 May 1785, in *Papers of Thomas Jefferson*, 8:160 (see n. 2).
42. "A Candid Address to the Freemen of the State of Rhode Island on the Subject of the Approaching Election," in *History of the American Presidential Elections, 1789–1968*, ed. Arthur M. Schlesinger, Jr. (New York: Chelsea, 1971) 1:140.
43. *Philadelphia Gazette of the United States*, September 23, 1800, 3.
44. *Philadelphia Aurora*, September 24, 1800, 2.
45. Thomas Jefferson to John Holmes, 22 April 1820, Library of Congress transcript, http://www.loc.gov/exhibits/jefferson/159.html (accessed May 10, 2009).

From Kin to Intruder: Cherokee Legal Attitudes toward People of African Descent in the Nineteenth Century

Fay A. Yarbrough

In the spring of 2007, the Cherokee Nation of Oklahoma grabbed national, and even international attention, with its March 3 decision to revoke the citizenship rights of 2,800 Cherokee freedmen—that is, the descendants of slaves owned by members of the Cherokee Nation during the nineteenth century—by stipulating that individuals be able to trace ancestry to the "by blood" rolls to claim Cherokee citizenship, in effect instituting a blood requirement for citizenship.[1] Of course, such rhetoric ignores the fact that some of the freedmen do in fact have Cherokee ancestry, but because of the policies of the Dawes Commission, their ancestors were placed on the "freedmen" rolls instead of the "by blood" rolls in the late nineteenth and early twentieth centuries.[2] The Dawes Commission, created by a federal act, enumerated the membership of various nations in the rolls, and each member then received a parcel or allotment of land. This process of enrollment and allotment was part of a larger policy of dissolving tribal governments and ending communal land ownership in favor of individual land ownership among the Cherokee Indians and other tribal groups.[3] Only the "by blood" roll of the Dawes Commission's enumeration includes information about Cherokee blood quantum, a concept that traditional Cherokee practice did not recognize.[4] Some indigenous groups use these lists or rolls as the basis for determining citizenship today.

As a part of the campaign to expel the freedmen descendants, Darren Buzzard circulated an electronic message that warned in part, "FOR OUR DAUGHTER[S] . . . FIGHT AGAINST THE INFILTRATION," a message that invoked the old fear of interracial sex.[5] John Ketcher, a member of the Cherokee Nation, claimed that he never saw a black person until he was ten years old and was skeptical that the freedmen descendants are part of the Cherokee community. "I think they want some of the goodies that are coming our way," he said.[6] His views no doubt reflect those of many others in the Cherokee Nation. These responses and the decision to revoke freedmen citizenship came in reaction to the successful legal attempt by Lucy Allen, a descendant of a Cherokee freedman, to obtain voting rights for herself and other freedmen descendants and is part of a much longer legal struggle over the inclusion or exclusion of those freedmen descendants in the Cherokee citizenry, a struggle with roots in the late eighteenth- and early nineteenth-century Cherokee Nation.

Throughout the nineteenth century, American officials and indigenous populations negotiated terms for coexistence. Federal and state governments were unsure just what status indigenous nations held. The federal government continued to make treaties with indigenous groups as distinct sovereign nations but frequently disregarded such agreements over territorial boundaries, and states such as Georgia insisted that indigenous governments had no authority and passed laws that applied to indigenous persons despite the objections of native peoples. Indigenous groups such as the Cherokees, on the other hand, understood their national sovereignty to be complete and total. To emphasize this point, during the first third of the nineteenth century, Cherokee legislators wrote a constitution much like that of the United States Constitution. It included provisions defining membership in the Cherokee Nation, outlining voter eligibility, regulating property ownership, and establishing criminal behavior and punishment.

At the same time, Cherokee Indians formulated and formalized their own racial ideologies and attitudes toward individuals of African descent. Indian removal, the Civil War, the end of slavery, and the introduction of the population of freedmen combined to create a tumultuous century in the Cherokee Nation when ideas about race and racial difference crystallized. Several nineteenth-century legal cases delineate the evolution of Cherokee racial attitudes toward people of African descent. Together these cases demonstrate that official Cherokee policy on the treatment and inclusion of blacks in Cherokee society was in flux during the nineteenth century. Cherokee legal authorities incrementally developed a concept of Cherokee identity that excluded individuals of African descent. In my judgment, this change in Cherokee legal attitudes toward individuals of African descent reflects a shift in thinking about race and racial hierarchy in the larger Cherokee society. In other words, the legal variation in the treatment of blacks in the Cherokee Nation embodies the Cherokees' own changing self-perceptions within a larger racial hierarchy that included the United States.[7]

Early in this era, the principle social organizational feature of the Cherokee Nation was the concept of clan. Clan affiliations were paramount to legitimate membership in the Nation.[8] Clan ties offered basic protections to life through the operation of blood law and clan vengeance. Clans were responsible for the actions and punishments of their members and for the retribution of crimes committed against members. Within the Nation, the absence of a clan membership meant the absence of any individual rights that others were bound to respect. Matrilineal descent established clan membership, which meant that any children produced in unions between Cherokees belonged to the mother's clan. Likewise, the children of Cherokee women and non-Cherokee men also belonged to the mother's clan and therefore had an undeniable claim to membership in the Nation.[9] As long as the children remained in the Nation, Cherokee authorities recognized these children of mixed race as Cherokees culturally and racially. Cherokee men who produced children with European or African women, on the other hand, produced children with no clan identity.[10] Without clan ties, one could not claim legitimate or legal membership in the Nation. Upon marriage, clan affiliations did not change and Cherokee women maintained their own property rights. The husband moved into his wife's home, which was located in a female-centered household in which sisters or other female relatives shared labor.

Against this backdrop of political and social organization, a series of nineteenth-century legal decisions revealed the Cherokees' growing reluctance to include people of African descent as legal members of their society. Never static, Cherokee ideas about race and citizenship continued to develop throughout the period, as demonstrated by the stories of Molley and her descendants, Shoe Boots's petition for the citizenship of his children, and Susan and Lemuel Boles's civil suit. In the early nineteenth century, Cherokees granted those of African descent citizenship full legal and political rights. By 1827 Cherokee legislation recognized as legal citizens of the Nation only those persons of African descent who also had Cherokee ancestry, and it granted them only limited rights.[11] As the twentieth century approached, the Cherokee Nation made it clear through legislative action that those of African descent, to whatever degree, were different and unacceptable as full-fledged members of the Nation. The legal decisions produced by the Cherokee Nation expose the progressively diminishing status of people of African descent in Cherokee society.

Legal documents recorded the citizenship status of individuals, thus offering insight into which racial groups had access to legitimate membership in the Cherokee Nation. Documents such as court records might include legal testimony confirming or denying an individual's Cherokee citizenship. Cherokee citizens could file petitions with the National Council to extend citizenship to their children. Through marriage, Cherokee citizens conferred citizenship upon their American spouses. District clerks kept careful records of marriages between Cherokee citizens and foreigners to ensure that the Nation had accurate accounts of the "outsiders" who were lawful residents of the Nation. Legal documents provided official recognition of citizenship and could also mark changes in citizenship status. Cherokee authorities might also record their rejection of certain citizenship claims within legal documents. Cherokee officials' willingness to admit some groups of people to citizenship and reject others reflects larger societal racial attitudes. Thus the changing legal treatment of one racial group's claims to Cherokee citizenship provides a glimpse into the Cherokees' changing perceptions of race.

The story of the slave woman Molley and her descendants shows that during the late eighteenth century, Cherokees recognized individuals of African descent as potential equals and potentially legitimate members of the Nation. Molley arrived in the Cherokee Nation, located in North Carolina at this time, prior to the Revolutionary War under curious and tragic circumstances.[12] Molley's owner, Sam Dent, was a white slave trader who had married a Cherokee woman from the Deer clan. By "beating and otherwise mistreating" her during her pregnancy, Dent killed his wife. The Deer clan demanded satisfaction. One of the benefits of clan membership among the Cherokee Indians was protection; others would not harm you because they risked retaliation from your clan. Clan members were obliged to avenge each other's deaths.[13] Thus Dent knew that failure to appease the Deer clan would cost him his life. Dent bought Molley in neighboring Georgia and offered her to the Deer clan as a replacement for his dead wife. After a town council meeting, the clan agreed to accept Molley and her future descendants as members of the clan and Nation. Dent escaped retribution, and the Deer clan emancipated Molley and gave her a new name, Chickaua, symbolizing her integration into the clan and Nation.[14] Chickaua and her descendants became full and equal members of the Cherokee Nation with all the rights, privileges, and protections associated with clan affiliation despite her former status as a slave and her African ancestry.

Chickaua's story might not have been preserved were it not for nineteenth-century legal attempts to return Chickaua and her descendants to slavery. Apparently prior to the legal activity that precipitated a legal pronouncement of Chickaua's status, Chickaua and her sons had lived unmolested as members of the Cherokee Nation for more than four decades. By 1833 a woman named Molley Hightower claimed Chickaua and her descendants as her slave property. Hightower presented a bill of sale from Dent to Hightower's father, who had also been an Indian trader living near Chickaua and her children. Hightower petitioned the Nation to hand over her slave property. Members of the Deer clan responded swiftly. They asked the Cherokee Council to prevent the "oppressive and illegal wrong attempted to be practiced on our Brother and Sister by the Hightowers in carrying into slavery two of whom have ever been and considered native Cherokees."[15] Not only did the clan recognize Chickaua and her children as full and free members of the Cherokee Nation, but it defended her legal rights as a Cherokee and argued for her continued enjoyment of liberty and freedom. Other prominent Cherokees also certified the circumstances of Chickaua's entrance into the Nation and adoption into the tribe. They further stated that she and her sons, Isaac and Edward, had resided in the Nation and "enjoyed the liberty of freedom."[16] The Deer clan, as well as other members of the Cherokee Nation, supported Chickaua's claims to freedom and legitimate membership in the Nation.[17]

The legal documents also describe a transformation in identity: Molley, a slave of African ancestry, had become, in the language of other Deer clan members, a "native" Cherokee named Chickaua. In the late eighteenth century and early nineteenth century, then, Cherokees had a fairly broad understanding of Cherokee identity that was not tied to race or status but to clan. Molley, an enslaved black person, could become a Cherokee. Moreover, the Nation legally recognized Chickaua as a Cherokee because of her own behavior and her acceptance by the larger Cherokee community. In other words, there was a performative and public component to Cherokee identity.[18] Petitioners for Chickaua's freedom noted her and her children's continued residence in the Nation and other Cherokees' acknowledgment of the family's free status. This understanding of the nature of Cherokee identity implies that Cherokees had not yet developed a rigid racial ideology. Neither Chickaua's lack of Cherokee ancestry nor her African ancestry excluded her from Cherokee identity. Further, Chickaua could move not only from slavery to freedom but also from black to Cherokee.

Cherokee legal authorities' willingness to accept Chickaua as a citizen is more striking given the legal context of the period. By 1833 the Cherokee Nation had passed laws restricting the rights of Cherokees of African descent and prohibiting marriages between "Negro" slaves and Indians or whites.[19] The passage of racially specific legislation highlights the legal contestation of the place of blacks in the Nation. Instead of extrapolating from this new legislation that Chickaua and her descendants were not entitled to Cherokee citizenship and returning them to slavery, the Nation chose to honor older notions of what it meant to be Cherokee. The Deer clan placed clan membership above considerations of race or status and pushed National authorities to respect the clan's determination of Chickaua's legitimacy as a Cherokee citizen. Chickaua's legal citizenship then conferred legitimacy upon her sons because clan membership was fixed matrilineally. Anthropologist Circe Sturm stresses how Chickaua's story exemplifies the strength of the matrilineal clan system, because Chickaua's sons retained Cherokee citizenship through her, yet the racial implications of the story are equally dramatic.[20] By the end of the nineteenth century, African ancestry would preclude full and equal membership in the Nation, but Chickaua's Cherokee citizenship reflects the values of a period that viewed Cherokee identity as less racially rigid.

While Chickaua and her sons lived inconspicuously, unaware of the looming questions of citizenship and legitimacy, a Cherokee named Shoe Boots was demonstrating that a shift in attitudes toward people of African ancestry was already taking place. On October 10, 1824, Shoe Boots, a "full-blood" Cherokee warrior, wrote a letter to the council chiefs requesting that the Nation recognize his children's freedom and grant them Cherokee citizenship.[21] The mother of Shoe Boots's children was not a Cherokee; this fact alone excluded the children from Cherokee citizenship. Shoe Boots's children had no clan identity and thus were not Cherokees. Race and status further impeded official Cherokee recognition of Shoe Boots's children. Shoe Boots had these children with his black slave Danell.[22] He confessed in his letter, "Being in possession of some black people and being crest, in my affections, I debased myself, and took one of my black women by the name of Danell, by her I have had three children . . ."[23] Shoe Boots feared the "bone of my bone and flesh of my flesh" would be distributed among his friends and family as property according to his will should he die. He could not bear the thought that his children and grandchildren might remain in perpetual bondage. Shoe Boots asked not only for the freedom of Elizabeth, John, and Polly but also for their official recognition as Cherokee citizens.

Shoe Boots's personal history represents the tremendous potential for and variety of interracial relationships that could exist in the Cherokee Nation. His experiences serve as a kind of metaphor for the larger Cherokee Nation's interactions with the black and white populations of the United States. Shoe Boots had initially, married a Cherokee woman. While still married, Shoe Boots captured Clarinda Ellington, a young white girl, in a 1792 raid. Shoe Boots and his wife kept Clarinda as a slave. When Shoe Boots's Cherokee wife died some years later, he and Clarinda married and soon had several children. By this time, Clarinda's family had located her and tried unsuccessfully to convince her to return home to Kentucky. Finally, through some subterfuge and despite promises that the stay would be brief, Clarinda returned to her family with her children in 1804. Shoe Boots would never see his children by Clarinda again. A few years later, Shoe Boots began a relationship with his black slave and with her had several children who became the subject of this petition.[24]

The council's response reveals its ambivalence toward Shoe Boots's relationship with his slave woman and his part-black children. The Cherokee National Council agreed to grant freedom to Shoe Boots's three children by a "slave" and recognize "their inheritance to the Cherokee Country."[25] The council claimed to have "no objections" regarding the petition; however, the council included a reprimand with the bestowal of freedom and Cherokee citizenship, which belied this assertion. The Council admonished Shoe Boots to "cease begetting any more Children by his said Slave Woman."[26] The council was willing to accept Shoe Boots's part-black children as free citizens of the Nation, but council members clearly felt uncomfortable with the sexual relationship that had produced the children. Council members officially warned Shoe Boots that his relationship with a black slave woman was improper and inappropriate. Shoe Boots had violated taboos of sex between individuals of different status and different races. Less than a week after the council responded favorably to Shoe Boots's request, and perhaps in reaction to the issues Shoe Boots's request magnified, the council passed legislation that prevented intermarriage between Negro slaves and Indians or whites.[27] Some consideration was likely given to Shoe Boots and his request because of his prominence in the Nation. He was a venerated war hero with ties to the Cherokee elite.[28] Thus despite objections to Shoe Boots's relationship with Danell, the council admitted his children to the citizenry on November 6, 1824.

The council's admonition did not, however, deter Shoe Boots from continuing his relationship with Danell, by whom he had two more sons. Some scholars have described Shoe Boots's years-long relationship with Danell as a marriage or a common-law marriage.[29] Several contemporary observers also described Shoe Boots and Danell and husband and wife and testified that Shoe Boots claimed Danell as his wife when questioned about the matter.[30] Shoe Boots, however, did not refer to Danell as his wife in his petition and even described his relationship with her as "debas[ing]." Further, by 1824 Cherokee lawmakers had legally prohibited marriage between slaves and free people.[31] Perhaps what was most telling was that Shoe Boots had asked for freedom and Cherokee citizenship only for his children but not for Danell. After his death in 1829, his sisters inherited the two other children as slave property and petitioned the council to emancipate them and grant them Cherokee citizenship, but the council refused, citing its earlier warning to Shoe Boots. Again, Shoe Boots's sisters neglected to ask for the freedom or citizenship of their nieces' and nephews' mother. The two boys remained slave property, and eventually a white man claimed them as well as the rest of Shoe Boots's estate. The claimant removed the boys from the Nation and separated them from their Cherokee family. Shoe Boots's worst fears for his mixed African and Cherokee children were realized.

Shoe Boots's situation may have prompted further Cherokee legislation to clarify the limits of interracial marriage and the ability of people of African descent to access Cherokee citizenship. The 1824 statute that prohibited marriage between Indian or white free citizens and "Negro slaves" came less than one month after Shoe Boots's petition for citizenship for his children and may have been in direct response to this particular situation.[32] Shoe Boots's petition provided a legal record of his relationship with a slave woman. Further, legitimate children are produced by legitimate unions. Granting legitimacy to Shoe Boots's children had the potential to legitimize the union between Shoe Boots and Danell, at least legally. Cherokee legislators may have wanted to prevent even the appearance of permitting formal relationships between Cherokees and slaves. Without official sanctions or prohibitions against marriage between slaves and masters, outsiders might interpret long-standing sexual relationships between slaves and masters or slaves and free people as a form of common-law marriage. Sexual relations between masters and slaves remained a prerogative of the master class; lawmakers were careful not to circumscribe it, but the law specifically precluded the formalization or legalization of these relations as marriages. Lawmakers then had to explicate the rights of any children such informal relations might produce.

The council wrote a clause into the 1827 Cherokee Constitution excluding the children of Cherokee men and women "of the African race" from the citizenry.[33] Conversely, this clause did extend citizenship to "the children of Cherokee men and white women living in the Nation as man and wife." The Nation was expressing through legal activity the limits of Cherokee identity: individuals of African descent without an undeniable claim to clan membership could not be lawful Cherokees. On the other hand, Cherokee intermixture with whites was acceptable, even when such unions produced children without clan affiliations, since only the children of Cherokee women had clear claims to clan membership and Cherokee identity. Hence the Nation recognized as legitimate citizens only the part-black children of Cherokee women. Further, to underscore the undesirability of people of African descent in the citizenry, the council limited the political rights of these individuals.[34] In other words, the Cherokee Nation always recognized as citizens the children of Cherokee women and any non-Cherokee man, but restricted only the rights of children produced in unions between Cherokee women and men of African descent. The Constitutional clause points to the resiliency and strength of matrilineal clan affiliations while acknowledging that racial considerations could nullify rights based on clan relationships.

Having admitted three of Shoe Boots's mixed-race children to citizenship, the Cherokee Nation thereby extended the full protections of Cherokee citizenship to each of them. As free citizens of the Cherokee Nation, Elizabeth and Polly conferred status and citizenship upon their own children. It is unclear if John would have been able to confer citizenship upon any children by a non-Cherokee spouse, given the persistent importance of matrilineal ties to the Nation and his African ancestry. Through Elizabeth and Polly, Shoe Boots had several free, part-black granddaughters. When white slave hunters captured Shoe Boots's granddaughters near Fort Gibson some years after Removal, the Cherokee Nation sent two Cherokees, Charles Landrum and Pigeon Half Breed, to recover the women. Landrum and Half Breed pursued the slave catchers to Missouri and freed the women. The council then reimbursed Landrum and Half Breed for the expenses incurred while acting on behalf of Cherokee citizens.[35] Although the council may have only reluctantly granted Shoe Boots's children citizenship, the council's commitment to these part-African Cherokee citizens was complete. Cherokee legal authorities recognized and honored its commitment to these citizens despite their African ancestry. By the end of the nineteenth century, however, lawmakers would exclude people from Cherokee citizenship because of their African ancestry.

The period between Shoe Boots's petition to the Cherokee Council and the Boleses' civil suit was turbulent and traumatic for the Cherokee Nation. In the late 1830s the federal government forcibly removed southeastern Indians (perhaps most famously the Cherokee Indians) to Indian Territory, located in the northeastern corner of present-day Oklahoma. The Cherokees experienced a period of civil war once they arrived as competing factions struggled for control of the new government. The new constitution and government the Cherokees finally established was largely a carryover from their pre-Removal institutions. Many Cherokee slaveholders had traveled to the Indian Territory with their human property and expected the new Cherokee legislature to continue to protect their property rights and access to slave labor.

Then the American Civil War and intervention by the federal government forced the Cherokee Nation to accept its former slaves as citizens. In 1863 the Cherokee Nation abolished slavery and soon after "declared that the freedmen had no rights of privileges as Cherokee citizens and were to be treated as members of other nations or communities."[36] Cherokee authorities denied that the ex-slaves had any legitimate claims to Cherokee citizenship, in part because they had no clan and kinship ties to the Nation. The American federal government disagreed and insisted in the Treaty of 1866 that the Cherokees must grant citizenship to their former slaves.[37] Without the federal government's stipulations, the freedmen would have been in the odd position of lacking citizenship in any nation; the Fourteenth Amendment did not apply to Cherokee freedmen because they resided in a separate sovereign nation and their former owners were not U.S. citizens. The Cherokees, however, promptly attempted to reduce the number of freedmen who could claim Cherokee citizenship by refusing to grant citizenship to freedmen who returned to the Nation after a six-month time limit. By 1886 Cherokee lawmakers further refined the limits of citizenship for the freedmen by restricting their access to annuity funds and land. The freedmen only had "civil, political, and personal rights," according to National authorities.[38]

The introduction of the large population of freedmen into the Nation through the Civil War and Emancipation did not soften Cherokee racial attitudes toward people of African descent, but it did serve to sharpen the boundaries of Cherokee citizenship.[39] Prior to 1866, the vast majority of people of African descent living in the Nation were slaves whose status effectively barred them from legal citizenship. A small number of people of African descent did claim Cherokee citizenship; however, as previously stated, during the antebellum era, the Cherokee legislature passed laws to limit this population and further constrained the rights of those legitimate claimants. The existence of a racialized system of slavery permitted the Nation to avoid questions and controversy about the legal place of blacks in Cherokee society. When the federal government forced the Nation to grant Cherokee citizenship to the ex-slaves, Cherokee authorities faced a dilemma. The Nation responded by legally redefining Cherokee citizenship and creating new categories of citizens with varying legal rights. Cherokee freedmen could not be full citizens who exercised all the rights of native Cherokee citizens—that is, citizens by birth and not adoption. By establishing a racial hierarchy of citizenship, the Nation preserved racial distinctions between native Cherokees and blacks. An examination of marriage between Cherokee freedmen and African Americans, as permitted by the suit between the Boleses and George Vann, demonstrates the constrained nature of citizenship for freedmen members of the Nation.

The heart of the legal dispute between the Boleses and Vann was property rights. Each party claimed ownership of an improvement. Cherokee Nation law did not permit the outright individual ownership of land by citizens, but provided for citizens to access as much land as they could improve. The law required only that improvements to the land, such as homes, barns, fences, or plowed fields, remain one-fourth of a mile away from a neighbor's improvement.[40] The Boleses, asserted that they had made an improvement and then sold it to Vann. After Vann was unable to make payments on the property, the Boleses reclaimed the improvement and the surrounding land. Vann maintained that he had purchased the improvement for Bill Vann; thus Bill Vann was responsible for paying the Boleses and was the only party who could return the improvements to them. Bill Vann continued to live on the land and make further improvements. He then rented improvements to yet another party, Fox McCaleb. The litigants asked the Cherokee courts to sort through the tangle of conflicting claims to determine the lawful owners of the improvements. The plaintiffs' race and citizenship status further complicated efforts to ascertain the legal ownership of the improvements: Susan Boles was a Cherokee freedwoman married to Lemuel Boles, a U.S. citizen of African descent.[41] At this time the nature of the citizenship rights of Cherokee freedpeople was often unclear and increasingly under attack by Cherokee Nation authorities.

In a common legal tactic of the day, the defendant made a motion to dismiss the case for want of jurisdiction. The outcome of the motion was contingent on Lemuel Boles's citizenship status. Vann claimed that Lemuel Boles was not a citizen of the Cherokee Nation but a colored foreigner who had married a colored Cherokee citizen, by virtue of the Treaty of 1866, which admitted freedmen to the Cherokee citizenry.[42] Vann further claimed that the National Council and decisions by the U.S. Department of the Interior declared that any foreigner who married a freedman citizen of the Cherokee Nation was not entitled to the same rights as natives of the Cherokee Nation. Cherokee courts had no legal authority to handle civil cases involving U.S. citizens, and Vann was proposing that Lemuel Boles, despite his marriage to a Cherokee citizen, remained a U.S. citizen. Lemuel Boles would reply that he had obtained a license to marry in the Saline District from the court clerk and that the district judge had officiated the ceremony.[43]

Lemuel Boles asserted the validity of his case and his right to Cherokee citizenship based on Cherokee law, specifically statutes governing intermarriage. He based his claim on his marriage to a Cherokee freedwoman. His response rested on two basic principles: a Cherokee freedperson could confer citizenship on an American spouse, and the Cherokee Nation's legal recognition of Susan Boles as a Cherokee citizen and of their marriage implied an acceptance of his right to Cherokee citizenship. The statement that a district judge had officiated their union implies a judicial recognition by deed of the marriage's validity and Lemuel Boles's claim to citizenship. His action of bringing the suit to court suggests that he viewed his citizenship status as secure; that is, he had no idea that his legal activity was putting his Cherokee citizenship in jeopardy. Why would Lemuel Boles purposely attract legal scrutiny to himself unless he thought his Cherokee citizenship was unquestionable? While Vann's position depended on legal decisions espoused by Cherokee legislative and judicial bodies and U.S. federal decrees, Lemuel Boles's argument relied on the functioning of the law. In other words, Vann would reason that laws overrode the daily actions of clerks and judges; conversely, Lemuel Boles would privilege legal action—the behavior of district clerks and judges—over official statements of Cherokee policy.

Court documents corroborate Lemuel Boles's account of legal marriage. On August 2, 1868, the Saline district clerk recorded a marriage between Leonard Bowles and Susan Vann.[44] This legal entry matched Lemuel Boles's statement that his marriage to Susan had taken place in the Saline District. Interestingly, the clerk noted that Susan Vann was a colored citizen of the Cherokee Nation and referred to Bowles only as a U.S. citizen. Given the existing laws prohibiting marriage between blacks and any other racial group, the clerk may have assumed that the groom's race was implied when the bride's race was indicated.[45] The clerk also recorded that Bowles's petition was signed by several Cherokees affirming his good character and responsible nature and supporting his efforts to marry into the Cherokee Nation. These petitions usually attested to the worthiness of the applicant for Cherokee citizenship. The form of this marriage entry matched the numerous surrounding entries for marriages between white men and Cherokee women "by blood" (a term the district clerks often included in the official marriage record).[46] All white grooms were entitled to Cherokee citizenship through marriage. The Saline district clerk, at least, treated Bowles as though Bowles was gaining Cherokee citizenship through his marriage to Susan Vann.

That Susan Boles and the defendant once shared the same surname suggests a possible familial connection that further complicates the operation of law and marriage in this legal battle. The legal documents do not divulge George Vann's race; however, he is most likely the George Vann that appears in the 1880 Census of the Cherokee Nation.[47] The 1880 census identifies this George Vann as "adopted colored" and aged thirty. The legal documents connected to the case indicate that George Vann was forty-eight years old in 1898, which fits the age of the Vann listed in the census.[48] Moreover, the 1880 census indicates that Vann was living in Cooweescoowee District in 1880, as were Susan Boles and several other participants in the court case: Eli Vann, Dank (Duck) Vann, and William Vann, Jr. Each of these participants is also listed as "adopted colored" in the 1880 census.[49] The Vann family had been an extremely prominent slaveholding family prior to the Civil War. By 1835 Joseph Vann of Hamilton County, Tennessee, was the largest slaveholder in the Cherokee Nation with 110 slaves. Several other Vann family members owned slaves as well.[50] Many Cherokees moved to the Indian Territory with their slave property along with other moveable goods.[51] Perhaps Susan Boles and Vann were cousins, or, at least, slaves owned by the same family. If not, their being of similar age and living in the same vicinity suggests the two likely knew each other prior to their legal suit.

A George Vann appeared on the petition in support of Leonard Bowles's request for a license to marry Susan Vann, a detail that adds more complexity to the question of the relationship between Vann and Susan Boles. The petition clearly affirmed Bowles's "good moral Character" and "industrious habits" before recommending "him to become a citizen of the Cherokee Nation by marrying a col'd woman and citizen of the said Dist. and Nation."[52] In 1868 at least, Vann did believe that colored citizens of the Cherokee Nation could confer Cherokee citizenship on American spouses. Vann publicly and legally supported Susan Vann's right to confer citizenship on a noncitizen spouse, as well as Bowles's right to claim Cherokee citizenship. The petition included the signatures of seven Cherokee citizens, as required by the 1855 law governing marriage between Cherokee citizens and noncitizens.[53] Two other signatories were also Vanns; Susan Vann likely turned to male relatives and close familial connections (perhaps a former owner) to obtain the needed signatures for the petition.[54] Presumably the signers had also thought that colored Cherokee citizens could bestow citizenship on noncitizens through marriage. The district clerk duly recorded the petition and marriage as though he concurred with this interpretation of the intermarriage law's applicability to colored Cherokee citizens. In light of Vann's former support of the union between Susan Vann and Leonard Bowles, as well as Bowles's claims to Cherokee citizenship, Vann's subsequent attempts to have Bowles's Cherokee citizenship repudiated and Susan Vann's citizenship rights circumscribed are all the more startling.

Faced with strong evidence favoring both parties, the Cherokee court took no action, choosing instead to hold the case under advisement. Finding Lemuel Boles's citizenship status difficult to determine, the court opted for more time to review the evidence. The court's indecision is surprising in light of the legal opinion issued in 1884 that prohibited Cherokee district clerks from issuing marriage licenses to colored noncitizens to marry Cherokee freedmen, because the intermarriage laws applied only to whites.[55] Similarly, in 1885 A. B. Upshaw, acting commissioner of Indian affairs, informed the U.S. Indian agent that he thought Cherokee freedwomen could not confer citizenship on colored noncitizens.[56] Hence the decision in 1889 should have been clear: Lemuel Boles's marriage to Susan Vann did not entitle him to Cherokee citizenship. Consequently, Lemuel Boles could not bring the suit to Cherokee court, and his claim against Vann should have been dismissed. Cherokee lawmakers, who had been forced by federal officials to extend citizenship to the Cherokee freedmen, should have seen the case between the Boleses and Vann as an opportunity to limit the rights of Cherokee freedmen and prevent the introduction of more colored people to the citizenry. Instead, the court hesitated to take a stand.

To be fair to the judge of the Cooweescoowee court, Cherokee law itself was unclear about the status of marriages involving colored people, be they Cherokee citizens or American citizens. Although anti-amalgamation statutes, which prevented marriages between colored persons and Indians or whites, remained on the books, regardless of the citizenship of the parties, district clerks recorded several marriages between Cherokee women and African American men.[57] The marriages followed the same legal procedures outlined by the intermarriage act that regulated marriages between Cherokee women and white men. The district clerks did not question the legality of these interracial unions despite the existence of the anti-amalgamation laws. Were these unions actually legal? Did the anti-amalgamation law invalidate the actions of district clerks? No record of legal decisions to answer these questions remains. The lack of legal documentation on the prosecution of individuals for violating bans on particular kinds of interracial marriage suggests that interracial couples usually lived unmolested by the authorities. It is likely that only when the couple attracted negative legal attention did Cherokee officials step in and make a determination about the citizenship of the intermarried colored men. While the threat was not always executed, Cherokee/black couples lived with the constant danger of having their unions dissolved by legal officials.

Cherokee legal attitudes toward intermarried African American men were contradictory at best. Intermarried colored men would complain that they received fewer rights than intermarried white men in the Cherokee Nation.[58] But according to Cherokee legal pronouncements, intermarried African American men should not have received any rights from the Cherokee government. Further, Cherokee lawmakers had also defended their right to expel intermarried African American men from the Cherokee Nation.[59] Clearly, Cherokee officials permitted intermarried African American men to remain in the Cherokee Nation despite the restrictions and granted them some civil rights, which indicates a degree of legal recognition and legitimacy of interracial marriages between Cherokee women and African American men. The juxtaposition of the presence of intermarried African American men in the Cherokee Nation and the legal injunctions preventing marriages between African American men and Cherokee women or between African Americans and Cherokee freedmen exposes an incongruity between Cherokee legal thought and Cherokee legal behavior.

The ambiguity in Cherokee law between proclamation and enforcement, between what the law stated and what lawmakers prosecuted, created space for interracial unions within the Cherokee Nation. This space existed between whom the Cherokee freedmen could legally marry and whom they actually married without legal repercussions. Obviously, Cherokee freedmen could marry other Cherokee freedmen with no question about the validity of the union. Marriages between Cherokee citizens who were not of the same race, however, provoked uncertainty. These marriages would not introduce new outsiders to the citizenry and further stretch Cherokee land and monetary resources, yet the anti-amalgamation act would make these marriages illegal. This fact adds further support to the argument that the anti-amalgamation law was about race and racial hierarchies, not citizenship and status. Cherokee authorities, however, rarely pursued the noncitizen or the Cherokee violators of the anti-amalgamation statute. No unions between Cherokees "by blood" and Cherokee freedmen were noted in the marriage records of this study, perhaps because freedmen were not in danger of being removed from the Nation as intruders; thus the legal certification such marriages by district clerks was unnecessary.

In the end, the court's indecision about Lemuel Boles's Cherokee citizenship paid off: Lemuel Boles withdrew from the suit, rendering questions about his citizenship status moot. The case, instead, reappears in the court docket with Susan Boles as the plaintiff.[60] In effect, Lemuel Boles circumvented any controversy about his right to Cherokee citizenship by removing himself from the actionable parties. Neither legal authorities nor the defendant questioned Susan Boles's Cherokee citizenship. With Susan Boles as the plaintiff, the case could not be thrown out on jurisdictional grounds. Perhaps the couple was also concerned that the court's decision might be negative and lead to Lemuel Boles's removal from the Cherokee Nation. In any case, the court avoided being forced into making a determination, and the Boleses could proceed with their suit. Curiously, Vann also bowed out of the legal proceedings by having his name "stricken from the suit as being a party not interested in the case."[61] Thus, the original parties in the suit receded from it as the litigants to appear only as witnesses in the later proceedings.

The legal testimony from the suit also exposes how active women were in the public activities of initiating legal actions in Cherokee courts and conducting business transactions. Susan Boles replaced her husband as the plaintiff in the case; she also bought and sold property independently and, apparently, with little interference from him. Susan Boles had directed her husband to make the initial purchase of the land in dispute for her. He testified that after the purchase, he told his wife and Vann "to make their own trades."[62] Susan Boles then made her own demands to Vann for payment and staked her own claim on the property by settling her daughter Jennie and some household goods on it.[63] When the various male parties to the transaction had questions or concerns about the property or wanted to negotiate a sale, they sought Susan Boles, not her husband, to make arrangements. Clearly Susan Boles was recognized as a property holder in her own right. This recognition, along with the challenged citizenship claim of Lemuel Boles, might also explain her substitution as the plaintiff in this court case. Perhaps her autonomy was a holdover from the traditional matrilineal roots of Cherokee society. Susan Boles may have grown up seeing Cherokee women control and manage their own property and continued in that example. Her standing as a legitimate Cherokee citizen through whom Lemuel Boles gained access to Cherokee citizenship may have also reinforced her ability to control property. With the citizenship status of intermarried African Americans in doubt, Cherokee citizens may have seen Susan Boles as the lawful arbiter of property in the Boles family.

Cherokee lawmakers neglected to reexamine the laws regulating intermarriage after the introduction of a sizeable number of colored people into the citizenry. Court action could reveal the inadequacies of marriage laws regarding colored people because of the conflicts between the United States and the Cherokee Nation concerning legal jurisdiction. Cherokee courts had to establish the citizenship of the parties involved before trying a case. Thus, individuals of questionable citizenship who operated outside the purview of Cherokee authorities, especially those who claimed Cherokee citizenship through a Cherokee citizen spouse, faced legal scrutiny when called into court as witnesses or defendants or when they initiated legal action themselves. The courtroom, then, became a site that magnified the weaknesses of the intermarriage laws in dealing with the black population.

The Boles/Vann court case laid bare the tangled relationship between citizenship, marriage, and the judicial system, as well as inter- and intraracial tensions within the Cherokee Nation. Courtrooms in the Cherokee Nation served as stages for human dramas, which also had national ramifications. The Cherokee legislature had not adequately addressed the question of freedmen's rights. Further, Cherokees had not created a social space for the new colored members of their citizenry. Instead, the Cherokees continued to operate with a social hierarchy based on slavery in an environment where slavery no longer existed. To regulate what it meant to be Cherokee, Cherokee lawmakers created a gradated scale of legal citizenship in which different people possessed different rights. The Cherokee Nation's inability to adjust to the realities of their post-Emancipation society would lead to controversies over the membership criteria in the Cherokee Nation that persist even today.

The Cherokees were not the only population forced to reconsider the legal rights and place of blacks in their society; the American South faced a similar dilemma. The Civil War and the Emancipation obliged Southerners, both Indian and white, to address the terms for inclusion of the newly emancipated black population. What legal rights should be accorded to the ex-slaves? Should they be citizens? Should they be enfranchised? Slave or free status, which had formerly been closely tied to race, could no longer organize society. White Southerners and Cherokees began to rethink the meanings of race in light of the sizeable number of newly freed blacks in their midst.[64] The federal government set the terms for the readmission of the Southern states to the Union and similarly dictated the conditions for the continued existence of the Confederate Indian nations. Federal authorities compelled white Southerners and Indian governments to recognize the citizenship rights of the ex-slaves. Freedpeople in the Cherokee Nation and Southern states could now vote, hold office, and own property, and these rights were to be respected by fellow citizens. Both societies responded to the political and social uncertainties wrought by Emancipation by firming racial divisions through legislative action.

The Cherokee Nation and the American South followed a similar trajectory in their treatment of blacks in the post-Emancipation era. The turmoil and upheaval of the Civil War and its resolution prompted a period of uncertainty during which the racial and social order were unclear. Without a system of racially based slavery to designate the place of blacks, Southern and Cherokee legislatures turned to writing laws to clearly delineate the boundaries of acceptable black behavior and limit the citizenship rights of blacks. Southern states enacted Jim Crow laws that carefully divided public spaces racially, prohibited interracial marriage, and disenfranchised black voters. Likewise, Cherokee lawmakers continued to outlaw interracial marriages and meticulously circumscribed freedmen's citizenship rights. As the 1880 Boles/Vann civil case demonstrates, Cherokee freedmen lacked the full rights of Cherokee citizenship and, from the perspective of most Cherokees, were unwelcome additions to the Nation. The developments in Cherokee attitudes about blacks after the Civil War, then, were symptomatic of larger changes taking place in the minds of many Southerners.

The stories of Chickaua née Molley, Shoe Boots, and Lemuel Boles form a timeline that illustrates changes in Cherokee thinking about race and racial difference during the nineteenth century. Cherokees' acceptance of Chickaua and her sons as citizens and their vigorous defense of their right to full and continued citizenship represent a moment when Cherokee notions of race and identity were less rigid than they would become. Given her adoption by the Deer clan, neither Chickaua's African ancestry nor her lack of Cherokee blood barred her from legitimate citizenship. Further, clan adoption apparently made her Cherokee by nationality; that is, she became a Cherokee woman capable of bearing Cherokee children. Shoe Boots's petition and the ensuing treatment of his African-Cherokee descendants mark a moment of transition: authorities disapproved of the relationship that produced the African-Cherokee children but recognized a father's desire to protect his children. Having granted full citizenship to Elizabeth, Polly, and John, the Nation honored commitments to these new citizens and their offspring. Authorities, however, promptly wrote legislation to prevent a reoccurrence of Shoe Boots's situation. And by the time Lemuel Boles's civil suit reached the Cherokee court system, racial lines had hardened and the Nation had firmly established a policy of excluding people of African descent from full citizenship.

The manner in which Cherokee courts and legislators dealt with questions of citizenship for people of African descent divulges how lawmakers saw Cherokee society racially. Judicial and legislative action throughout the nineteenth century reflects a changing social plan or vision of the kind of society Cherokee lawmakers were hoping to create. At the beginning of the century, Cherokees could admit a person of African descent to full citizenship, and clan considerations could supercede racial ones. By the end of the nineteenth century, however, Cherokees clearly found colored people unacceptable as full citizens and refused to allow any more blacks than necessary into the citizenry. Moreover, the Cherokee legislature refused to allow those black citizens equal rights with Cherokee citizens by birth or even intermarried white citizens. Whites, on the other hand, were acceptable—and even desirable—as Cherokee citizens. The Cherokee government established a clear and specific plan for the inclusion of whites in the Nation. Of course, intermarried whites did not have all the rights and privileges of Cherokees "by blood." Cherokee authorities were careful to circumscribe the citizenship rights of white people entering the Nation. As expected, this configuration of the Cherokee social hierarchy placed Cherokees on top, gave whites a nod for their ability to uproot the Nation by allotting them intermediary status in the social order, and firmly consigned blacks to the bottom.

The legal dispositions of these cases not only articulated conceptions of race, however; the decisions also expose a society reimagining gender relationships. Traditionally, matrilineally determined clan affiliations had organized Cherokee society. As the producers of legitimate members of the Nation, Cherokee women had enormous power; clan memberships obtained by birth through a Cherokee woman were undeniable. Adoption, such as that of the Georgian slave Molley, was the only means by which outsiders could become members of the Nation.[65] Older practices of adoption generally admitted war captives—most frequently women and children—to clan membership, which may help explain the Deer clan's ready acceptance of Molley.[66] The matrilineally determined clan membership granted Molley's, now Chickaua, sons, Isaac and Edward, Cherokee citizenship. Shoe Boots's petition, however, suggests that the importance of matrilineally determined clan affiliations were on the decline in the nineteenth century. The Cherokee Council admitted his children with Danell, a slave woman without clan membership, to the citizenry because of their patrilineal connection to the Cherokee Nation, despite of their lack of a matrilineal one. Finally, the Boleses' suit demonstrates that women continued to serve as legal actors in Cherokee courts and hints at the perseverance of older practices of landholding among Cherokee women despite the other dramatic changes that took place during the period.

Throughout the nineteenth century, the Nation aligned itself ever more closely with the white race and adopted a racial ideology that focused on the difference between black and nonblack instead of white and nonwhite. The Cherokees focused on the similarities among all nonblack peoples, drawing connections between themselves and white Americans in a bid to gain recognition of Cherokee sovereignty. Cherokee lawmakers hoped that by demonstrating to white Americans what Cherokee Indians and white Americans had in common, the federal government would finally honor treaty obligations and respect the physical and jurisdictional boundaries of the Cherokee Nation. Cherokees rightly realized that white Americans had an alarming tendency to lump American Indians and people of African descent into one larger nonwhite category, and they sought to sharply distinguish themselves from people of African descent through laws, such as limiting the citizenship rights of people of African descent and prohibiting marriage with people of African descent, and behavior, such as owning slaves.

In 1907, in a move commensurate with the large Indian population in the area, the state Constitution of Oklahoma defined race in a manner that gave special consideration to American Indians and continued to place blacks in a position of inferiority:

> *Wherever in this Constitution and laws of this state, the word or words "colored" or "colored race," "negro" or "negro race," are used, the same shall be construed to mean or apply to all persons of African descent. The term "white race" shall include all other persons.*[67]

The state then enacted Jim Crow legislation, separating "whites" and "colored" people in public spaces. The racial policy of barring blacks from legitimate citizenship, of according them lesser status, however, was not in place at the outset of the nineteenth century. Throughout the century, Cherokees developed and adjusted official policy to determine the place of people of African descent in Cherokee society. People of African descent, even those without clan ties to the Nation, could and did become full and equal members of the Nation. By the end of the century, however, there was little space for individuals of African descent in the Cherokee Nation. And recent events in the Cherokee Nation, such as the expulsion of the Cherokee freedmen from the citizenry, suggest that there continues to be little space for people of African descent there in the twenty-first century.

Notes

1. Associated Press, "Cherokees Vote to Limit Tribal Membership," March 4, 2007, Washington Post Company, http://www.washingtonpost.com/wp-dyn/content/article/2007/03/03/AR2007/3/31705_pf (accessed March 7, 2007; site now discontinued).
2. Celia E. Naylor, *African Cherokees in Indian Territory: From Chattel to Citizens* (Chapel Hill: University of North Carolina Press, 2008), 179–87.
3. For more on the allotment process, see Angie Debo, "The Surrender to the United States" and "The Dissolution of Tribal Interests," in *The Rise and Fall of the Choctaw Republic* (Norman: University of Oklahoma Press, 1934), chapters 11 (245–68) and 12 (269–90); Debo, "Breaking Up the Reservations," in *A History of the Indians of the United States* (Norman: University of Oklahoma Press, 1970), chapter 16 (299–315); Naylor, "Cherokee Freedpeople's Struggle for Recognition and Rights during Reconstruction" and "Contested Common Ground: Landownership, Race Politics, and Segregation on the Eve of Statehood," in *African Cherokees in Indian Territory*, chapters 5 (155–78) and 6 (179–200) (see n. 2); and Circe Sturm, *Blood Politics: Race, Culture, and Identity in the Cherokee Nation of Oklahoma* (Berkeley: University of California Press, 2002), 78–81.
4. For more discussion of blood and blood quantum in the Cherokee Nation, see Fay A. Yarbrough, *Race and the Cherokee Nation: Sovereignty in the Nineteenth Century* (Philadelphia: University of Pennsylvania Press, 2008), 42–43.
5. Ellen Knickmeyer, "Cherokee Nation to Vote on Expelling Slaves' Descendants," March 3, 2007, Washington Post Company, http://www.washingpost.com/wp-dyn/content/article/2007/03/02/AR2007030201647_pf (accessed March 7, 2007; site now discontinued).
6. Adam Geller, "Past and Future Collide in Fight over Cherokee Identity," February 10, 2007, *USA Today*, http://www.usatoday.com/news/nation/2007-02-10-cherokeefight_x.htm (accessed April 23, 2009).
7. For a more thorough treatment of racial ideology in the nineteenth-century Cherokee Nation, see Yarbrough, *Race and the Cherokee Nation* (see n. 4).
8. For discussions on matrilineal descent in Cherokee Society, see Theda Perdue, *Cherokee Women: Gender and Culture Change, 1700–1835* (Lincoln: University of Nebraska Press, 1998), especially 81–83 and "Constructing Gender," chapter 1 (17–40); "Defining Community," chapter 2 (41–64); and "Trade," chapter 3 (65–85); Perdue, *Slavery and the Evolution of Cherokee Society: 1540–1866* (Knoxville: University of Tennessee Press, 1979), 9; John Phillip Reid, *A Law of Blood: The Primitive Law of the Cherokee Nation* (New York: New York University Press, 1970), 113–22; J. Leitch Wright, *Only Land They Knew: The Tragic Story of the American Indians in the Old South* (New York: Free Press, 1981), 235; Henry Thompson Malone, *Cherokees of the Old South: A People in Transition* (Athens: University of Georgia Press, 1956), 17; Sturm, *Blood Politics*, 28 (see n. 3); Katja May, *African Americans and Native Americans in the Creek and Cherokee Nations, 1830s to 1920s: Collision and Collusion* (New York: Garland Publishing, 1996), 33–34; and Rennard Strickland, *Fire and the Spirits: Cherokee Law from Clan to Court* (Norman: University of Oklahoma Press, 1975), 49–50.
9. Sturm, *Blood Politics*, 31 (see n. 3).
10. Sarah Hill, *Weaving New Worlds: Southeastern Cherokee Women and Their Basketry* (Chapel Hill: University of North Carolina Press, 1997), 94. Hill states

that intermarriage with European or African women produced children with no clan identity.

11. Cherokee Nation, *Laws of the Cherokee Nation, Adopted by the Council at Various Periods, Printed for the Benefit of the Nation*, The Constitution and Laws of the American Indian Tribes, vol. 5 (Wilmington, DE: Scholarly Resources, 1973), 120. This volume contains two sections. The second section is "The Constitution and Laws of the Cherokee Nation: Passed at Tahlequah, Cherokee Nation, 1839–1851." In this example from the 1827 Cherokee Constitution, individuals of either "negro or mulatto parentage, either by the father or mother side" were barred from holding public office in the Nation. The children of black women and Cherokee men were also excluded from the citizenry.
12. *Record Book of the Proceedings of the Supreme Court of the Cherokee Nation*, Cherokee Collection, reel 1, box 3, folder 10: 0984–0987, Tennessee State Library and Archives, Nashville, TN. For a brief summary of the case, see Sturm, *Blood Politics*, 57–58 (see n. 3). Shoe Boots's story can also be found in Marion L. Starkey, *The Cherokee Nation* (New York: Alfred A. Knopf, 1946), 18–19; and Tiya Miles, *Ties That Bind: The Story of an Afro-Cherokee Family in Slavery and Freedom* (Berkeley: University of California Press, 2005).
13. Reid's *A Law of Blood* addresses the meaning of clan in the Cherokee Nation (see n. 8). For more discussion of the importance of clans to governance in the Cherokee Nation, see also Strickland, *Fire and the Spirits* (see n. 8).
14. *Record Book*, 0985–0986 (see n. 12). In Sturm's account in *Blood Politics* (see n. 3), Molley's new Cherokee name is Chickaune. I read the evidence as listing Molley as Chickaua née Molley.
15. Ibid., 0985.
16. Ibid., 0986.
17. Interestingly, the documents contain no mention of Molley's spouse and if she had one, nor do they name the father of her children. One might presume that the children's father was a Cherokee, but we cannot be sure.
18. For more on race as a performative component to identity, see Ariela J. Gross, "Litigating Whiteness: Trials of Racial Determination in the Nineteenth Century South," *Yale Law Journal* 108 (1998): 109–88.
19. Cherokee Nation, *Laws of the Cherokee Nation*, 38, 120 (see n. 11). Page 38 contains the 1824 act preventing intermarriage with Negro slaves.
20. Sturm, *Blood Politics*, 58 (see n. 3).
21. Cherokee Nation Papers, roll 46, folder 6508, Oklahoma Historical Society, Oklahoma City.
22. Starkey, *Cherokee Nation*, 18–19 (see n. 12); and Strickland, *Fire and the Spirits*, 99–100 (see n. 8). These authors have referred to this slave woman as Lucy. The original letter written by Shoe Boots refers to a woman named Danell or possibly Daull. Anthropologist Circe Sturm in *Blood Politics* (see n. 3) also interprets the slave woman's name as starting with a *D* and uses the name Daull. In other documents, individuals testify that the slave woman's name was Dolly. See M1650, 1896 Citizen Application #4422, Oklahoma Historical Society, Oklahoma City. Miles refers to this woman as Doll in her extended study of this relationship between Shoe Boots and his slave property in *Ties That Bind* (see n. 12). See especially appendix C for the issue of names.

23. Cherokee Nation Papers (see n. 21).
24. The details of Shoe Boots's relationship with Clarinda Ellington are in William Stephens's 1896 petition for Cherokee citizenship. He was the grandson of Shoe Boots and Clarinda Ellington and gave a detailed account of the circumstances for the marriage between Ellington and Shoe Boots and her subsequent return to Kentucky. Stephens filed his petition on August 1, 1896, with the U.S. Commissioner to the Five Civilized Tribes and included testimony from several individuals. I obtained copies of the petition from Tressie Nealy, volunteer archivist at the Oklahoma Historical Society, Oklahoma City. Other scholars have also summarized Shoe Boots's situation based on these documents. See especially Miles, *Ties That Bind* (see n. 12); and Sturm, *Blood Politics*, 58–61 (see n. 3). Briefer accounts that confirm the situation but differ in some details can be found in Strickland, *Fire and the Spirits*, 99–100 (see n. 8); and May, *Collision and Collusion*, 45 (see n. 8).
25. Cherokee Nation Papers (see n. 21). The word *slave* is underlined in the original response, which is found with the request.
26. Ibid.
27. Cherokee Nation, *Laws of the Cherokee Nation*, 38 (see n. 11).
28. See James Mooney, *Myths of the Cherokee* (Washington, DC: Government Printing Office, 1902), 394; and Starkey, *Cherokee Nation*, 24–25, 57 (see n. 12).
29. May, *Collision and Collusion*, 63 (see n. 8); Miles, *Ties that Bind*, 183–85 (see n. 12); and Sturm, *Blood Politics*, 60 (see n. 3).
30. See the testimony from William Shoeboots's application for Cherokee citizenship, M1650, roll 51, 1896 Citizenship Application #4422, Oklahoma Historical Society, Oklahoma City. William was a descendant of the Shoe Boots and the slave woman Danell. The informants, contemporaries of Shoe Boots and Danell who were asked to testify about the couple's relationship, relied on behavior to determine the nature of the relationship. For instance, Nathaniel Fish stated, "What (caused) me to know that it was his wife while were at the table eating Shoeboots and his wife eat on one side of the table I eat on the other side." William Shoeboots's application included Fish's statements in an effort to demonstrate his legitimate claim to Cherokee citizenship.
31. Cherokee Nation, *Laws of the Cherokee Nation*, 38 (see n. 11).
32. Ibid.
33. Ibid., 120, art. 3, sec. 4.
34. Ibid. According to the Cherokee Constitution of 1827, the children of Cherokee men and black women had no rights to Cherokee citizenship, and the children of Cherokee women and black men could not hold political office in the Cherokee Nation despite being citizens.
35. May, *Collision and Collusion*, 63 (see n. 8). See also Sturm, *Blood Politics*, 60–61 (see n. 3). One cannot help but consider the implications of Half Breed's last name and what an interesting convergence of race his name represents.
36. Daniel F. Littlefield, Jr., *The Cherokee Freedmen: From Emancipation to American Citizenship* (Westport, CT: Greenwood Press, 1978), 16.
37. *Treaty of 1866*, art. 9. Similarly, the federal government demanded that the Confederate South recognize the citizenship rights of the freed slaves by passing the Fourteenth and Fifteenth Amendments to the U.S. Constitution. The federal

government, in this sense, treated the Cherokees and other indigenous nations in the same manner as it did the other members of the Confederacy.

38. Cherokee Nation, *Constitution and Laws of the Cherokee Nation, Published by an Act of the National Council 1892*, The Constitutions and Laws of the American Indian Tribes, vol. 10 (Wilmington, DE: Scholarly Resources, 1973), 371–72.
39. For a more detailed discussion of the effects of the Civil War and Emancipation on the lives of people of African descent in the Cherokee Nation, see Naylor, *African Cherokees in Indian Territory* (see n. 2).
40. See Cherokee Nation, *Laws of the Cherokee Nation*, 40–41; and "The Constitution and Laws of the Cherokee Nation: Passed at Tahlequah, Cherokee Nation, 1839–1851, 29 (see n. 11). Cherokee Nation, *Compiled Laws of the Cherokee Nation, Published by Authority of the National Council*, The Constitutions and Laws of the American Indian Tribes, vol. 9 (Wilmington, DE: Scholarly Resources, 1973), 309–310 . The Cherokee legislature reiterated the same provisions related to improvements and use of the public domain time and again in 1824, 1839, and 1870.
41. Cherokee National Records Microfilm Series, roll CHN 117, vol. 247, 83–104, Oklahoma Historical Society, Oklahoma City.
42. Within the Cherokee Nation, the term "colored" specifically referred to people of African descent. Hence someone of only Cherokee ancestry could not be a person of color, but a person of Cherokee and African ancestry would be colored.
43. Cherokee National Records Microfilm Series, roll CHN 29, vol. 49, 226–27, Oklahoma Historical Society, Oklahoma City.
44. Cherokee National Records Microfilm Series, roll CHN 44, vol. 157A, 15–16, Oklahoma Historical Society, Oklahoma City. I am confident that the slight variation in Boles's name is due to clerk error and Lemuel and Susan Boles's illiteracy. Clerks often spelled names they way they sounded, and these differences in spelling led to the same pronunciation in this case. The consistency with which Susan Vann's name appears, the confirmation of the location of the marriage in Saline District, and the small number of colored individuals who appear in court documents all support my contention that Lenard Boles, Lemuel Boles, and Leonard Bowles are in fact the same person. There were also few colored Susan Vanns listed on the enrollment records and only one was of the appropriate age to have married and participated in legal action in the Cherokee courts.
45. Littlefield, *Cherokee Freedmen*, 17 (see n. 36). See also Peter Wallenstein, "Native Americans Are White, African Americans Are Not: Racial Identity, Marriage, Inheritance, and the Law in Oklahoma, 1907–1967," *Journal of the West* 39 (2000): 57. The volumes that compile the laws of the Cherokee Nation contain no repeal of the anti-amalgamation statute, even after the Civil War. A law prohibiting intermarriage between people of African descent and all other groups was written into the Oklahoma state constitution in 1907.
46. See Cherokee National Records, 7–14, 18–176B (see n. 44). The marriage documents often included the designation "Cherokee 'by blood'" to describe the bride, groom, or signers of the petition in support of legal marriage.
47. See the detailed enumeration of the 1880 Cherokee Nation Census available at the Oklahoma Historical Society, Oklahoma City. George Vann is listed as number 3036 in the Cooweescoowee District.

48. Cherokee National Records, 87 (see n. 41).
49. See the detailed enumeration of the 1880 Cherokee Nation Census (see n. 47). Eli Vann is listed as number 3051, Dank Vann as number 3028, and William Vann, Jr., as number 2128. All were residents of Cooweescoowee District and of the same age cohort.
50. See William G. McLoughlin and Walter H. Conser, Jr., "The Cherokees in Transition: A Statistical Analysis of the Federal Cherokee Census of 1835," *Journal of American History* 64 (1977): 683, 696.
51. R. Halliburton, Jr., *Red Over Black: Black Slavery Among the Cherokee Indians* (Westport, CT: Greenwood Press, 1977), 59. See also *Indian Pioneer History*, vol. 23, microfilm, 40. The Indian Pioneer History records were collected by the Works Progress Administration and are available at the Oklahoma Historical Society, Oklahoma City.
52. Cherokee National Records, 15–16 (see n. 44).
53. Cherokee Nation, *Laws of the Cherokee Nation, Passed During the Years 1839–1867, Compiled by Authority of the National Council*, The Constitutions and Laws of the American Indian Tribes, vol. 6 (Wilmington, DE: Scholarly Resources, 1973), 104–105.
54. The other signers included Charley Rogers, J. Vann, Je [illegible] Vann, Dennis Murrell, Jonathon Webber, and David Rowe. J. Vann might have been Joseph Vann, a former large slaveholder in the Cherokee Nation. Incidentally, David Rowe was also the district court judge who performed the Boleses' marriage ceremony.
55. Cherokee National Records Microfilm Series, roll CHN 85, vol. T, Marriages, 1880–1889, Oklahoma Historical Society, Oklahoma City. This volume contains no page numbers. The decision is included in a March 15, 1884, letter from Illinois district clerk J. Thornton to acting principal Chief R. Bunch.
56. Ibid. The opinion is included in a September 4, 1885, letter from A. B. Upshaw to U.S. Indian Agent John W. Tufts.
57. Murray R. Wickett, *Contested Territory: Whites, Native Americans, and African Americans in Oklahoma, 1865–1907* (Baton Rouge: Louisiana State University Press, 2000), 35; and Littlefield, *Cherokee Freedmen*, 17 (see n. 36). For examples of marriages legally recorded between Cherokee women and colored men, see Cherokee National Records Microfilm Series, roll CHN 24, vol. 19C, 16, 22 Oklahoma Historical Society, Oklahoma City.
58. May, *Collision and Collusion*, 199 (see n. 8).
59. Wickett, *Contested Territory*, 35–36 (see n. 57).
60. Cherokee National Records, 315, 366 (see n. 43).
61. Ibid., 315.
62. Cherokee National Records, 90 (see n. 41).
63. Ibid., 88, 91, 98.
64. For discussions on the Reconstruction and its effects on Southern society, see Edward L. Ayers, *Southern Crossing: A History of the American South, 1877–1906* (Oxford: Oxford University Press, 1995); Eric Foner, *A Short History of Reconstruction* (New York: Harper & Row, 1990); C. Vann Woodward, *Origins of the New South, 1877–1913* (Baton Rouge: Louisiana State University Press, 1997); Woodward, *The Strange Career of Jim Crow* (New York: Oxford University

Press, 1957), especially the introduction; Joel Williamson, *A Rage for Order: Black-White Relations in the American South Since Emancipation* (New York: Oxford University Press, 1986); Ayers, *The Promise of the New South: Life After Reconstruction* (Oxford: Oxford University Press 1992); Glenda Gilmore, *Gender and Jim Crow: Women and the Politics of White Supremacy in North Carolina, 1896–1920* (Chapel Hill: University of North Carolina Press, 1996); and Grace Elizabeth Hale, *Making Whiteness: The Culture of Segregation in the South, 1890–1940* (New York: Pantheon Books, 1998).

65. May, *Collision and Collusion*, 34 (see n. 8); and Theda Perdue, *"Mixed Blood" Indians: Racial Construction in the Early South* (Athens: University of Georgia Press, 2003), 9.
66. Perdue, *"Mixed Blood" Indians*, 8 (see n. 65).
67. Wallenstein, "Native Americans Are White," 57 (see n. 45).

The Last Repatriationist: The Career of Earnest Sevier Cox

John P. Jackson, Jr.
Andrew S. Winston

Introduction

"Race," the infamous British anatomist Robert Knox declared in 1850, "is everything: literature, science, art; in a word, civilization, depends on it."[1] Civilization for Knox was biological, resting as it did on a substrate of race. Outspoken in his belief that the colonial regime of the British Empire was a biological mistake, Knox insisted, "If we now inquire into the history of the Anglo-Saxon colonies we shall find that on quitting his native soil the Saxon loses all respect for it. He is totally devoid of the weakness called patriotism. His adopted land becomes his fatherland." Nor, Knox warned, should the Saxon colonizer believe that civilization could be brought to those unlike the Saxon himself, for "all races of men equal to a social condition, which in courtesy we may call civilization, will, I think, obey *the law* if made *by themselves;* law and government are identical and nearly synonymous terms. If in accordance with *their race*, the law is obeyed cheerfully."[2] The law was biological, and biology dictated that the races not intermingle in the same nation. For Knox, only by complete and utter geographical separation could racial integrity be maintained.

Views such as Knox's resonated with the racist writers of the United States. In his magisterial history of the ideology of white supremacy in the American South, Joel Williamson identified a stream of "radical" thought that maintained that the Negro had no place in the United States, even under conditions of racial segregation.[3] These views were not confined to Southerners. New York Nordicist Madison Grant, for instance, noted that geographical proximity was a biological disaster: "Where two distinct species are located side by side, history and biology teach that but one of two things can happen; either one race drives the other out, as the Americans exterminated the Indians and as the Negroes are now replacing the whites in various parts of the south; or else they amalgamate and form a population of race bastards in which the lower type ultimately preponderates."[4] His colleague Lothrop Stoddard agreed, writing against both "doctrinaire liberals," who believed in racial equality, and "doctrinaire imperialists, who maintained the equally imperceptible right of their particular nation to 'vital expansion' regardless of injuries thereby inflicted upon other nations."[5] Segregation, as practiced in the American South, and colonialism, as practiced all over the world, were for these thinkers biologically unsound legal regimes. They believed that the races must be kept geographically isolated for each to live in a biologically suitable civilization.

The uniting of anticolonialism and antisegregation may look odd to twenty-first-century eyes because "the ideology of imperialism did inspire the architects of segregation in the United States and South Africa."[6] Moreover, the rejection of racism in the latter half of the twentieth century was concomitant with the rise of anticolonial movements that made common cause with racial egalitarianism. Richard King, among others, has traced how "after World War II, just as the colonial empires were on the verge of collapse; ideologies of universal equality, independence, and democratic self-rule were gathering support in the West."[7]

The writings of Knox, Grant, and Stoddard make clear that the link between anticolonialism and antiracism is contingent rather than necessary. Throughout the history of racist thought, an alternative thread has been an anticolonialism and antisegregation version of scientific racism. In the United States, one way this thread of thought manifested itself was in the ongoing attempt to completely remove African Americans from American soil. These attempts began in 1817 with the foundation of the American Colonialization Society based on the idea that "blacks represented a danger to the order, stability, and progress of white civilization. It also represents irrefutable evidence of the whites' persisting belief that there were vast differences between themselves and the Negroes."[8] The repatriation of African Americans "back" to Africa waxed and waned throughout the nineteenth century.

Throughout his long life, Earnest Sevier Cox (1880–1965) was the most important white spokesman for the repatriation of African Americans to Africa in the twentieth century. Cox had enjoyed some political and legal success in the 1920s by championing the Virginia Racial Integrity Act, which strengthened the state's antimiscegenation laws. His views on the necessity of repatriating African Americans were driven by an extreme scientific racism that was intellectually respectable in the 1920s but out of fashion after World War II. Rediscovered by white Southerners after the *Brown* decision in 1954 in the era of "massive resistance," Cox's views on race remained unchanged throughout the twentieth century; his views on the racial interpretation of history were the same in the 1960s as they were in the 1920s. Despite his rigid position, however, Cox's views were widely accepted by a variety of audiences in the postwar United States. He appealed to segregationists though he was no segregationist. He appealed to Black Nationalists though he was a staunch white supremacist. He was most influential in the furthest reaches of the far right, where he found those who agreed with his antisemitic worldview. Cox could address multiple audiences by creating arguments that appealed to his far-right fellow travelers, traditional Southern segregationists, and the nascent Black Nationalists. When courting Black Nationalists, Cox would mute his overt racism and adopt tones of racial separation and anticolonialism.

Background of Earnest Sevier Cox

Born in Tennessee in 1880, Cox was a Methodist preacher as a young man. After studying at Vanderbilt University, but not receiving a degree, Cox attended the University of Chicago in 1906, pursuing graduate studies in sociology. At Chicago, Cox focused on the "Negro question," arguing in a series of papers that African Americans were, as a race, inferior to whites, and their brutality was held in check only by white governance. Writing to his sister in 1906 regarding the Atlanta race riot, wherein white mobs attacked and killed large sectors of Atlanta's African American population, Cox declared his eagerness to join in the violence, noting that the only true solution was for returning the "black savage to his native jungle."[9]

While at Chicago, Cox took a course from Frederick Starr, who had just returned from a tour of Africa. With Starr's encouragement, and additional support from Charles Henderson and W. I. Thomas, Cox set off on his own tour of that continent in 1910. By 1914 Cox had returned from Africa and had embarked for South America. Promoting himself as an "ethnological expert," Cox claimed he was preparing to write a book on interracial contacts. Upon his return to the United States, Cox worked on his manuscript while assisting Mississippi senator James K. Vardaman, who read a portion of Cox's in-progress manuscript into the *Congressional Record*.[10]

Cox soon developed into a minor public figure. When he resettled briefly in his native Tennessee, he lectured to "substantial audiences," complete with lantern slides of his travels. He received newspaper coverage in Knoxville, Nashville, and Washington, D.C.[11] In late 1915 the *New York Times* carried a short piece describing Cox's six-year journey to study the "negro problem" after his graduate work in political science, sociology, and ethnology. His fundamental view that the colored races produced no civilization was quoted without comment.[12] This piece attracted the attention of W. E. B. Du Bois, who noted in the *Crisis* that Cox's "absolutely false" position could only be maintained by claiming that Egyptians, Abyssinians, East Indians, and most Africans are "white." Du Bois took Cox to exemplify the way in which history, for some, consisted of "lies agreed upon."[13]

After a brief stint of military service during World War I, Cox settled in Richmond, Virginia, where he sold real estate to support his writing. Despite having completed his manuscript and receiving some encouragement from Grant, Cox was unable to find a publisher. Finally, Cox paid for the printing of *White America* himself.[14] The book was a crude retelling of Grant's *The Passing of the Great Race,* and the similarities were not accidental. As Jonathan Spiro, Grant's biographer, made clear, Grant tutored Cox in the findings of modern biology and eugenics, guiding the "half-educated" Cox's manuscript toward coherence.[15]

Cox made his initial assumptions clear by declaring in the first chapter that "scientific research has done much toward establishing the following propositions:

> 1. *The white race has founded all civilizations.*
> 2. *The white race remaining white has not lost civilization.*
> 3. *The white race, become a hybrid has not retained civilization.*[16]

He argued that only those who did not really live with the Negro could believe in racial equality. "The teachings of the whites who live apart from the negro have placed great emphasis upon environment, rather than upon race and heredity," Cox wrote, "whilst those whites who live in daily contact with the colored races are agreed that there is a difference between the white and colored which cannot be bridged by present environment." Cox situated his work firmly in the latest scientific thought by couching his calls for racial purity in the language of eugenics. "A race devoid of creative genius is an unfit type," he declared, "and is a matter of concern for the eugenist. Those who seek to maintain the white race in its purity within the United States are in harmony with the ideals of eugenics."[17]

It was history, rather than the young science of genetics, however, that served as evidence for Cox's claims. The bulk of *White America* was a review of civilizations that have perished because of miscegenation, including Egypt, India, China, Mexico, and Peru. "The intensity of civilization," Cox concluded, "is in inverse ratio to the numerical proportion of the negro in the populations of the various cultural centers."[18] From a contemporary perspective, the reliance on history might suggest that Cox's views were not scientific, since we usually think of genetics as providing the foundations for "scientific" racism. Historically, however, the separation of "biology" from "culture" and "history" is a relatively recent phenomenon.[19] In the context of Cox's times, the use of history was considered entirely appropriate to support an essentially biological argument. Even Comte Arthur de Gobineau's notorious racial theory of history was rooted in the pre-Darwinian biology and natural history of his age. For both Gobineau and Cox, biology and inherited qualities—not politics—explained the fate of nations: political history was the playing out of natural history. The careful analysis of conquest and degeneration—the history of contact among peoples—provided the essential evidence for this view.[20] The biological view of history continued as a powerful tradition within the writings of racial anthropologists of Grant's circle, which included Cox. Like Houston Stewart Chamberlain, Gobineau, and Grant, Cox argued that race was the driving force behind all historical development, and thus racial history should guide social action.[21] The only true solution to the menace of the Negro was complete separation through the physical removal of Negroes from the country.

As William Tucker has shown, *White America* was well received by a number of mainstream academics and scientists, such as psychologist William McDougall and sociologist E. A. Ross. When the *Eugenical News* suggested in 1924 that the removal of Negroes would make Cox a greater savior of his country than George Washington, the editor, noted eugenicist Harry Laughlin, invited Cox to speak at Harvard. Cox was treated as an important scientific authority by Dr. Walter Plecker, a Virginia official who rewrote the state antimiscegenation law in 1924.[22] *White America* was incorporated into the biology curriculum at the University of Virginia by Professor Ivey Lewis, head of the biology department and later dean of Arts and Sciences. Lewis invited Cox to address his class, and they continued to correspond through the 1940s.[23]

In the repatriationist cause, Cox also found allies in the Universal Negro Improvement Association (UNIA), founded in 1914 by Marcus Garvey, a Jamaican publicist and organizer who came to the United States in 1915 to meet with one of his heroes, Booker T. Washington. Like Washington, Garvey preached racial self-help with a focus on economic advancement rather than civil or political rights. Garvey's UNIA was dedicated to black self-help, including building a black economy that would eventually fund the creation of a homeland for African Americans in Africa. Like Edward Blyden and other Black Nationalist leaders, Garvey did not believe that salvation could come from white society. As Garvey's organization grew in popularity in the 1920s, he came under increasing attack from Du Bois, the National Association for the Advancement of Colored People (NAACP), and other racial progressives fighting for integration and political and civil rights for African Americans.[24]

By the time that Garvey and Cox began corresponding, Garvey was imprisoned for mail fraud for financial improprieties surrounding his steamship line. Garvey viewed his imprisonment as a political act instigated by those within the civil rights movement who opposed his focus on racial pride. He had turned to white supremacists in part because he sought political allies against racial moderates such as the NAACP and other "mulattoes," whom he despised.[25] He and his wife, Amy Jacques Garvey, corresponded with Cox regularly, After Garvey's death, his wife continued her correspondence with Cox, who continued his support of Garvey's repatriation plan until Cox's own death in 1965.

After Garvey's expulsion from the United States in 1927 left his organization without its charismatic leader, the UNIA never fully recovered. At the dawn of the 1930s, however, Cox soon found those who claimed to be the heirs to Garvey's dreams: Black Nationalists who called for a return of their people to Africa. In particular, he stayed in close contact with the leader of the Peace Movement of Ethiopia (PME), Mittie Maud Lena Gordon, who had visited Garvey in Jamaica after his exile and built the PME from signatures gathered during Mississippi senator Theodore G. Bilbo's repatriation campaign of the 1930s. Unlike Garvey, however, these nationalists were less interested in creating the black economy necessary to support voluntary repatriation and turned instead to federal funding for such efforts. Cox and his allies found a sponsor for their repatriation efforts; Bilbo, having received a copy of *White America*, vowed to make Cox's dream a reality.[26] An extreme racist even by the standards of the time, Bilbo cared even less than Cox about black self-reliance.[27] As the 1930s drew to a close, Bilbo sponsored the Greater Liberia Bill, which called for monies to be dedicated to supporting voluntary black repatriation to Africa. With many Black Nationalists who supported the bill deeply suspicious of the overtly racist Bilbo, Cox became the pivotal figure in holding the fragile alliance together during the debates over the Bilbo proposal. It was a short-lived effort, however; with the invasion of Nazi Poland in 1939, the meager white political support for Bilbo's program vanished and the effort collapsed.[28]

World War II fostered a new militancy within the African American community for civil rights and full participation in American democracy. Intending to hold white Americans to their promises of democracy and freedom, many African Americans were not content accepting second-class citizenship during wartime or afterwards. However, optimism about America's intent to extend full citizenship to African Americans was not universal. Those to whom Garvey held the greatest appeal remained suspicious of claims of full integration into white society:

Nonelite African Americans had grown weary of the unfulfilled pledges of the government to do better. Members of the Ethiopian Pacific Movement, Inc., the Peace Movement of Ethiopia, and Pacific Movement of the Eastern

World, the Temple of Islam, the Moorish Science Temple of America, and other groups were unconvinced that military victories over the Japanese would result in any appreciable gains or changes in the status of African Americans in the United States. They preferred to cheer the reports of Japanese triumphs over Allied Forces. These groups were composed of people who were dispossessed of any hope that African Americans might have a better future in the United States. They argued that whatever changes might occur as a result of a Japanese takeover of America, black people would be no worse off.[29]

Cox saw himself as the white representative to these Black Nationalists who clung fast to Garvey's dream of a return to Africa to free themselves of the oppression of white America and its string of lies to people of color. Gordon, who had lost a son to white mob violence in Springfield, Illinois, and her followers, who well remembered the broken promises and violence that followed World War I, had no reason to trust white Americans, or their integrationist African American allies, who claimed that World War II would bring racial justice to the country.[30] Moreover, the PME shared with Cox an ideology of racial separatism. Like Cox, PME leaders preached that two races could never share a geographical space with one another.

With such views bringing unwanted attention to the PME during the war, the group soon came under FBI surveillance. One FBI informant reported to J. Edgar Hoover rather breathlessly that the "ignorant and backward" Negroes of the PME were "pro-Nazi," and that at one PME meeting a speaker delivered a "speech [that] was primarily a eulogy of Hitler, showing that Hitler was not hostile to the Negroes. He said that Hitler said he did not hate the Negroes, he only pitied them." The FBI concluded that "active propagandizing for Nazism is going on among Negroes."[31]

There were deep affinities between Cox's ideology and the PME's. Both Cox and the PME held that the legal regime of a nation depended on racial purity. In an interview with the FBI, Gordon's husband, William Green Gordon, said, "The motto of our organization is, we must separate the black race from the white race. The white boys are on the south side destroying our girls and even some white girls come down on the south side to associate with black boys. We are against this because we want the black race to keep pure. We prefer that that the light colored women of our race should marry the blackest man she can find and the light colored man should marry the blackest woman so that their children will be all black."[32] Cox held that Jews were responsible for racial degeneracy, while Gordon "expressed antisemitic ideas . . . by blaming the Jews for the sorry conditions of the negroes, particularly making reference to the housing difficulties experienced by colored people."[33]

It was not an alliance between Black Nationalists and the Nazis that was of the most concern to the FBI; rather, it was that the Black Nationalists had pro-Japanese sympathies. Black Nationalist admiration of the Japanese was not a new phenomenon that sprang up in the 1940s. Forty years earlier, Black Nationalists had pointed to the victory of the Japanese over the Russians in the Russo-Japanese war in 1904 as proof that the darker races could indeed be victorious over the white race.[34] The Constitution of the PME held that there was a "Confraternity among All Dark Races."[35] The FBI became concerned about the PME when informants reported of Gordon telling her followers that Pearl Harbor was a blow for the freedom of the dark races, the attack proved that the white man was losing power, and it was time for all blacks to return to Africa.[36]

In 1942 the FBI arrested Gordon and a number of Black Nationalist leaders in a series of raids in Chicago. Gordon was charged with advising her followers to resist the draft and possible sedition with PME's pro-Japanese statements.[37] The FBI's investigation into the PME eventually led them to Cox's door. In his interview with the FBI, Cox expressed dismay that his allies had been arrested and tried to assure the government that the PME was not pro-Japanese. The interviewing agent made clear Cox's interpretation of what had transpired:

> *[Cox] said that there is a sharp division of opinion between the "Amalgamationists" and the "separationists." The former group, represented by such negro organizations as the National Association of Colored People, believes in fighting to raise the negro to the level of the white man and the eventual fusion of the two races by miscegenation. The latter group, composed of such organizations as the Peace Movement of Ethiopia and the Universal Negro Improvement Association, based the solution of separation of the races by transportation of the negro to Africa. COX said that his study of the race problem in various countries of the world had convinced him that the leaving of the negroes in this country would eventually be to make mulattoes out of all of us.*[38]

Cox further suggested to the FBI "that some of the 'Amalgamationists' had tried to 'frame' Madam GORDON as they had framed MARCUS GARVEY when he was imprisoned for selling stock to finance the Black Star Steamship line, which he was organizing to transport negroes to Africa."[39] While he did not make his views completely known to the FBI, Cox believed that the "amalgamationists" of the NAACP and other groups were controlled by Jews. Not persuaded by Cox's conspiratorial suggestion, government officials found Gordon guilty of advising her followers to avoid the draft and gave her two years probation.[40] After the war, Gordon and Cox would stay in close contact, hoping to further the cause of repatriation and thereby save both the white and black races.

The "Jewish Threat" to Teutonic Unity

When the war drew to a close, Cox's campaign for repatriation had been underway for nearly thirty years. In the 1930s and 1940s he received substantial personal and modest financial encouragement from Wickliffe Draper, founder of the Pioneer Fund.[41] Cox also continued to enjoy the support of Laughlin, who headed the Eugenics Record Office until 1939. Whenever it seemed as if repatriation would finally come before Congress, however, some new barrier arose, and the lack of success must have been painful for Cox. Nevertheless, remaining persistent and optimistic, he was able to make new alliances during the late 1940s. The tireless Cox kept up his correspondence with senators, congressmen, and university presidents with little effect.[42] But a new book, published privately in 1951, helped propel Cox into the inner circle of racial extremists and neo-Nazis.

Providing a popularized history of the Nordic Race stretching more than four thousand years, Cox's *Teutonic Unity* outlined the achievements of Teutons and other Aryan branches, juxtaposing their preeminent position as the founders and "carriers" of Western culture with their tendency to make war with one another. He reiterated and only slightly updated Grant's vision of racial history, while his views on the biology of race were hardly changed from those he espoused in *White America*: Teutonic peoples created culture and technology; black peoples were unable to do so. Mixture of the two produced the lowering of the higher race and a decline in the numbers of the very top achievers who were necessary for the successful progress and continuation of the race. Most significantly, race mixing would eliminate the special characteristics of the race that had developed over thousands of years. Once lost by mixing, the noble traits of the Teutons could never be recovered. Those who threatened the race by miscegenation often did so, Cox argued, because of a biological defect of "atavistic sexuality." Cox took a further step in *Teutonic Unity*; an enthusiastic Christian evangelist in his early years, he now argued that Christianity, "a Judaic religion," was inimical to the interests and survival of the Teutonic race.

The book was warmly received and recommended by segregationists, neo-Nazis, and paleo-Nazis such as Joseph Goebbels's former propagandist, Johann von Leers.[43] Cox was judicious in his remarks about World War II and even referred to the "barbarities" under Hitler. But his vision for the unification of Teutonic strains, as well as alliance rather than conflict with

Slavs, appealed to the sensibilities of neo-Nazi leaders, who hoped to capture a broader postwar audience than the narrowly defined Nordics. If the science of racial separation and the repatriation solution were both clearly laid out, as Cox hoped he had done, why was progress so difficult?

Cox had already provided the answer in 1947: the Jews. In his review of Ruth Benedict and Gene Weltfish's pamphlet *Races of Mankind*, Cox identified the nefarious and conspiratorial work of Franz Boas. Boas's rejection of the traditional racial hierarchy, Cox believed, was part of a plan for Jewish domination, to be carried out with the help of non-Jews.[44] The plan was to first destroy race science and race consciousness and then encourage race mixing, which could no longer be forbidden on scientific grounds. The resultant weakening of the Nordic race would make Jewish domination possible. This was not an original idea and had often been joined to the foundational Nazi belief that Jews directed Communism as part of the plan for world control outlined in the *Protocols of the Learned Elders of Zion*.[45] To Cox, the decline of race science by the 1930s was a perversion that could only be explained by the "Boas Conspiracy." The success of Boas student Ashley Montagu, née Israel Ehrenberg, in creating an official UNESCO statement on the equality of human races was further evidence of Jewish conspiratorial efforts.

Willis Carto and the Antisemitic Right

In the early 1950s Willis Carto, founder of the Liberty Lobby, and premier organizer of the racialist right and Holocaust denial in North America, was beginning his career. Although he would later break with most of his compatriots, for at least two decades he was the leading publisher of Nazi material and promoter of the ideas of Francis Parker Yockey, author of *Imperium* and the hero of the postwar Nazi underground. After Carto wrote to Cox in 1954 to commend him on *White America* and his review of *Races of Mankind*, their correspondence continued for the next decade. When Carto suggested that the "theory and science of Race, relatively new, becomes of staggering importance," Cox wrote back, emphasizing his work on repatriation and his cooperation with Negro Nationalist groups.[46] Their discussions over the subsequent years indicated a common vision of the scientific issues of heredity and race and the need for legal action. Like Cox, Carto aimed to bring about white racial preservation by legislation to protect segregation in all spheres. Up until 1954 the legal battle for Cox was over repatriation, but after *Brown*, the legal issues had a new and more urgent dimension: miscegenation would be the inevitable result of school desegregation. Those working energetically for desegregation would therefore be secretly aiming to bring about the destruction of the white race.

"I have long looked on the NAACP as an organization directed by Jews," Cox wrote to Archbishop Addison, a Black Nationalist leader.[47] Carto, Cox, and the members of the revived racialist right viewed the civil rights movement as "Jew controlled." They duly noted the evidence of Jewish involvement in the NAACP, particularly in the early years under the leadership of Joel Springarn. Cox wrote to *The White Sentinel*, a newsletter published by Gerald L. K. Smith's antisemitic crusade, that "the NAACP is a Jew organization using the Negro to get out of our concepts those which hinder the Jewish movement in our midst and get in concepts which will aid the Jew exploitation of our nation."[48] This vision was strengthened by the material that Cox received in his role as leader of white resistance. When he read Mrs. Lambert Schuyler's article "Why the Jew Is Hated" in the February 1959 issue of the antisemitic periodical *Truth Seeker*, Cox wrote to Mrs. Schuyler and offered his stock biological argument:

> *The Jew is a victim of his heredity and cannot do other than he does. With his throat slit in one country he goes to another country and repeats the conduct which caused him to be expelled from one country after another. The fact that he "acts like a Jew" in all countries shows that environment cannot change him.*[49]

As his communication with white supremacist groups increased, Cox received flyers and pamphlets announcing that integration, race mixing, and Communism were all part of the treasonous Jewish enterprise. In the Carto circle, the Jews blocked both the science of race and the legal remedies for race salvation. Jews were viewed as powerful enough to control Wall Street (through banking houses such as Kuhn & Loeb), international finance (through the House of Rothschild), and governments (by manipulating presidents, as in the case of Henry Morgenthau, Jr.'s alleged control over Franklin D. Roosevelt). If the United States government was already Jew-controlled, resulting in the threat of further civil rights legislation, then the need to vigorously lobby Congress was essential.

Although Carto's own position was extreme, he was a consummate coalition builder and attracted some to the Liberty Lobby who subscribed primarily to Jewish conspiracy theories, some whose primary interest was in race, and some who were extreme anti-Communists. Carto built an interlocking set of organizations with newsletters, magazines, books, a publishing house, a government lobbying organization, and even a pseudoacademic Holocaust denial body, the Institute for Historical Review.

The success of the Liberty Lobby depended on the enlistment of a wide range of respectable public figures such as Colonel Curtis Dall and novelist Taylor Caldwell. A steady stream of "news," analysis, historical works, and commentary were issued from this distinct intellectual community, thus providing a set of shared texts and shared meanings. Carto was able to gain supporters in Congress and provided them with more sanitized versions of his program. For Liberty Lobby insiders and readers of Carto's magazines, *Western Destiny* and *American Mercury*,[50] the program was clear: Carto promoted white racial superiority, international Jewish conspiracy theories, Franklin Roosevelt and Winston Churchill's responsibility for World War II, Holocaust denial, and extreme anti-Communism. Through the Noontide Press, Carto's publishing arm, Cox's work reached a wider audience, and Carto republished Cox's *White America* and *Teutonic Unity* as well as *Lincoln's Negro Policy*.

In the late 1950s Carto developed an alliance with Roger Pearson, an emerging leader of Nordic race consciousness in England. Along with Alastair Harper and Peter-Huxley Blythe, Pearson, working in India at the time, founded the Northern League for the friendship and unity of Nordic people.[51] The Northern League Statement of Aims revealed a deep concern of stopping the forces of "cosmopolitanism" and "degeneracy," which threatened Nordic biological and cultural existence. These thinly coded references to the Jewish threat constituted only slightly modernized versions of 1930s Nazi propaganda. The Northern League brought together figures such as Dr. Hans F. K. Günther, a leading race theorist during the Third Reich; ex-Waffen-SS officer and postwar neo-Nazi leader Arthur Ehrhardt; Robert Gayre, who built a coalition of academics to revitalize race science; and Charles Smith, American editor of the extreme antisemitic periodical *Truth Seeker*. Cox was an "honorary member," with membership card number three.

In 1955 Cox wrote to von Leers that "we should have a homecoming of many nations at the site of the battle of the Teutoberg Forest."[52] In *Teutonic Unity*, Cox told the story of Herman (Arminius), the German hero who defeated the Roman Legions in AD 9. Cox transformed Herman into a racial hero who saves the Teutonic race by preventing mixture with the Latin hereditary material of the already mongrelized Romans. In so doing, Herman rejected the path of his brother, who, to advance his career, took up arms for the Romans. As the Cox-Carto-Pearson cooperation developed, the opportunity to present this revised story emerged. Cox's repeated his suggestion of a meeting at Teutoberg Forest to both Pearson and Günther. Pearson wrote to Cox that, with his permission, the Northern League would "'steal' the idea if we may."[53] Eventually, a meeting, or "moot," was held at Detmold, Germany, in July 1959. The gathering included the laying of a wreath at the fifty-meter-high statue of Herman.

Although Cox did manage to visit the statue of Herman, he was too ill to read his speech at the Northern League moot, and it was read by Harper, who became a major figure in British Nationalist, racialist, and neo-Nazi circles.[54] At this gathering of old and new Nordicists from Europe and America, Cox was treated with great respect. With his cautionary tale of Herman's brother, Cox stressed the urgency for modern Nordics to react to the threats they faced, including threats from their own kind. As always, he pleaded for repatriation. In this speech, avoiding direct reference to the Jews, Cox instead referred to the communist idea of racial equality. For many in his audience, the two threats were equivalent. Cox's story of Herman was extraordinarily popular among racist groups of the late 1950s. As in Cox's early work, it was the use of history as evidence for his principles of race and race mixing that proved to be of broad appeal, particularly when set in a heroic narrative.

Cox's view of history as the primary source of scientific evidence on race was shared by an international and interdisciplinary group of academics that joined in Carto and Pearson's efforts to stop race mixing. This loosely organized community involved academics who were arguably "mainstream," such as psychologist Henry Garrett and geneticist R. Ruggles Gates, as well as those with more marginal reputations.[55] With the financial support of the Pioneer Fund, this coalition emerged in the 1960s as the International Association for the Advancement of Ethnology and Eugenics and was able to keep the embers of scientific racism alive despite a steady decline in followers. Although most of these scientists did not hope for a National Socialist revolution in America, many shared the sense of danger and the belief that the scientific truths regarding ineluctable racial differences were being suppressed. They treated Cox as a valuable intellectual ally in the coming struggle to preserve the white race.

Unending Hate

Welcomed as a hero among Hitler's ideological heirs, Cox was similarly embraced in the 1950s by white Southerners seeking to shore up the segregationist legal regimes of their state governments. In the wake of the *Brown* decision in 1954, the white South quickly declared that desegregation posed a threat to the "southern way of life" and vowed to block any attempts at racial intermingling through policies of "massive resistance." Constitutional historian Michael Klarman wrote that "*Brown* temporarily destroyed southern racial moderation" as all white Southern politicians moved "several notches to the right on racial issues."[56] Historian David Goldfield has underscored the importance of "respectable resistance" as many advocates of massive resistance portrayed themselves "as advocating preferable alternatives to race war."[57] On the local level, the Citizens' Councils were the most famous and largest of several respectable resistance organizations that sprang up throughout the South and held themselves as the alternative to the violence of quasi-military organizations such as the Ku Klux Klan. These organizations disagreed not with the Klan's goals of white supremacy but rather with the method of violence and intimidation; organizations like the Citizens' Councils would fight for the cause through reasoned discourse, legal battles, and the scientific truths of white supremacy.[58]

In this climate, many white Southerners rediscovered Cox's work. The organizer of the Florida Citizens' Council wrote to Cox with a list of books that the council wanted to distribute, noting that "I would certainly recommend your *White America* . . . were it possible" to obtain a sufficient quantity of the work.[59] Throughout the 1950s Cox received offers from the Citizens' Councils and similar groups to distribute his writings, "If you have a tract you would like us to distribute," wrote the Dallas Citizens' Council, "just send us one to read and tell us how many you can supply."[60]

With the new clamor for his writings after *Brown*, Cox had an opportunity to enlist more white support for the repatriationist cause. Within months of the *Brown* decision, he had published a pamphlet titled *Undending Hate*, attacking the decision.[61] However, Cox's work must be read as an expression of his particular racist ideology rather than as a typical pro-segregationist tract. Cox's Nordicism demanded that he view all events through the lens of race rather than politics, economics, or the law. Indeed, Cox's view was that when nations attempted to subjugate nature's racial demands to economic or political wishes, the result was a biological disaster.

Continuing his efforts to reach highly divergent audiences, Cox found himself in a curious position when called upon by segregationists. Having argued against *both* racial integration and continued racial segregation since the 1920s, in the 1950s he needed to find a way to convince white segregationists to embrace repatriation rather than continued segregation. An additional problem for him was that, because of his interest in maintaining his alliance with Black Nationalists, he could not preach the standard segregationist line on white supremacy without offending these allies. Cox's final work thus had to reach two different constituencies, neither of which trusted the other and neither of which was in agreement with everything he had to say.

Cox was to the right of the most extreme segregationists of the 1950s. In the 1920s he had written that segregation would never keep the United States racially pure. "The attainment of White America," he concluded in *White America*, "is not possible save by removing the Africans and excluding the Asiatics."[62] White Southerners who had actually read Cox's writings from the 1920s undoubtedly rejoiced at Cox's constant proclamations that Nordics were the pinnacle of racial evolution. However, his writings of the 1920s also made clear that Cox was no segregationist. Segregation, he argued, led to "mongrelizing the nation." He criticized white Southerners who continually made a place for the inferior race. Such an attitude, Cox claimed, was a danger to civilization since it led to miscegenation. He blamed the paradoxical position of "proclaiming the Negro a racial danger and yet clinging to him as an economic asset" on the white South's greed and demand for cheap labor.[63]

Segregation, Cox maintained, detracted from what should be white America's true goal: the repatriation of blacks back to Africa. Cox's struggle for repatriation thus continued despite Bilbo's death. In 1949, to Cox's surprise, he found that a repatriation bill had been introduced by Senator William Langer of North Dakota. Throughout the 1950s, Cox worked with Langer and a number of Black Nationalist groups to pass a repatriation bill.[64] Explaining that "my every breath has been given to this [repatriation] movement," Cox would not be distracted by wasting time with other efforts on behalf of white supremacy. "My efforts are given wholly to the one Cause of Negro colonialization" he insisted.[65]

Cox's second concern was allying himself with white supremacists without damaging his alliance with Black Nationalist groups. Cox had been walking this tightrope for decades (for example, by serving as the intermediate between the fanatical racist Bilbo and the remnants of Garvey's organization in the 1930s).[66] The key to the repatriationist movement, Cox believed, was not that blacks did not wish to move back to Africa but that there was a "lack of white support for it."[67] Hence if he could fashion a response that could convince whites to abandon segregation in favor of repatriation, he could advance his own cause.

In *Unending Hate*, Cox argued that the idea that interracial harmony could somehow be found in the U.S. Constitution was absurd. Cox was heir to a Nordicist strain of thought that tied democratic institutions to the particular racial heritage of Northern Europeans. In England, Nordicist doctrines began appearing in the sixteenth century, when the break with the Roman Catholic Church brought increased scholarly attention to the history of the Anglo-Saxon Church and, by extension, to other Anglo-

Saxon traditions. From the sixteenth century and into the early nineteenth century, English scholars extended these studies of Anglo-Saxon traditions by arguing that English common law and parliamentary democracy were in fact ancient Anglo-Saxon traditions rather than recent innovations.[68]

Many antebellum writers in the United States trumpeted the superiority of the Teutons. Tacitus's writings were taken as proof that the modern love of liberty could be traced to pre-Christian Germanic tribes. Many writers adopted these ideas into a theory of the "Teutonic origin" of democracy as proof against conservative critics who argued that democracy was an inherently unstable form of government. These writers argued that democracy originated in these Germanic tribes with their primitive parliaments and proto-representative government. According to this line of argument, the Teutonic tribes of Angles and Saxons brought this heritage to England and then across the Atlantic to the United States. Hence democracy was in some sense part of the ancient heritage of the Germanic people who settled in the United States. In the late nineteenth century, the theory of the Teutonic origins of democracy became racialized and the justification for democracy became minimal as the political ideal became part of the biological, rather than social, heritage of white Americans.[69]

Cox's allegiance to the Teutonic/racial origins of democracy was laid bare in his attack on *Brown*. "The Bill of Rights in our Constitution," he declared, "is not the product of the races of mankind. The Bill of Rights issued from Saxon and allied racial stocks which make up by far the greater portion of the nation's white population." To extend such concepts to those who are not racially equipped to deal them is to render rights and government "all but inoperative" because "the background of the Bill of Rights is tribal" and cannot be "used interracially."[70]

In its *Brown* decision, the Supreme Court cited social scientific evidence in its finding that segregation was psychologically damaging. In 1954 the U.S. Supreme Court ruled that racial segregation in public elementary and secondary schools was a violation of the equal protection clause of the Fourteenth Amendment of the Constitution. Chief Justice Earl Warren noted that "to separate [colored schoolchildren] from others of similar age and qualifications solely because of their race generates a feeling of inferiority as to their status in the community that may affect their hearts and minds in a way unlikely ever to be undone. . . . Whatever may have been the extent of the psychological knowledge at the time of *Plessy v. Ferguson*, this finding is amply supported by modern authority." Here, in the eleventh footnote of the opinion, the Court cited a number of psychological works to support its claim.[71]

The citation of scientific social evidence in *Brown* was the result of a decades-long campaign by the NAACP-Legal Defense and Education Fund to erode the separate-but-equal doctrine that had been enshrined in American constitutional law in the 1896 case of *Plessy v. Ferguson*. The NAACP's strategy, according to constitutional historian Mark Tushnet, was to transform an unfavorable case precedent into a favorable one by "pointing out anomalies in doctrine and identifying the inevitable failure of society's efforts to explain why unjust doctrines nonetheless were acceptable."[72]

For Cox, the citation provided a wedge for him to criticize *Brown* by pairing it with a most unlikely partner: *Dred Scott v. Sandford*:

> *In the Dred Scott case Chief Justice Taney considered the concept held by certain white men that the Negro was so inferior that it was a blessing for him to be enslaved by the white man. In the School Desegregation decision Chief Justice Warren imputed to the Negro a sense of inferiority when placed in Negro schools with Negro teachers. The Taney concept, above, of Negro inferiority has been abandoned by the higher civilized portions of mankind. Let us deal with the concept advanced by Warren.*[73]

Cox noted that Negroes were indeed inferior to whites in the United States, but claimed that was due to Negroes being forced to reside in a society for which they were unfit rather than argue that Negroes were innately inferior (though he undoubtedly continued to believe so). The problem was contact between the races rather than any inherent inferiority of the Negro. "When races are in contact there will be a race problem. Such problem cannot be solved save by the process of racial separation which will preserve the races."[74]

The separation suggested by Cox involved not just removing blacks from the United States but also withdrawing colonial powers from the African continent. Applying Warren's logic in *Brown* would, Cox maintained, "work havoc in Negro Africa. If the Negro in the United States will totter unless he is in reach of the white man's shoulder how can American Negroes champion the cause of their African brethren who wish to expel the white exploiters of Africa?" Cox noted that Black Nationalists he had worked with had championed the anticolonial movements afoot in Africa, and he himself had "long ago advocated that the white colonists be gradually withdrawn from Negro Africa."[75]

The Final Years of Earnest Sevier Cox

With the wide distribution of his works among neo-Nazis and the Citizens' Councils, Cox truly earned the title Carto had bestowed upon him: dean of American racists. There is no doubt that Cox clearly knew those he was dealing with, such as Commander George Lincoln Rockwell, founder of the American Nazi Party. Yet Cox also knew that he had put himself in a precarious position. In 1964, two years before his death, he had an awkward interchange with Carto and his associates. The managing editor of *Western Destiny* put Cox's name on the masthead as a contributing editor. Alarmed, Cox wrote emphatically to Carto that he could not be publicly identified with other groups. Although he admired *Western Destiny* and supported its efforts, he insisted that "my commitments to the Negro Nationalist movement prevented me from appearing in any form as a Contributing Editor."[76]

As his health deteriorated in 1964, Cox nevertheless kept up the fight for repatriation. He wrote to Matt Koehl, Rockwell's second in command at the American Nazi Party, "I work with Negro leaders who desire federal aid to settle on lands in Liberia held for immigrants in the United States. We will win this Cause, and all the Jews that are in Hell or out will not be able to eliminate the white race from the United States."[77] Such was the dual nature of Cox's alliances to both neo-Nazis and Black Nationalists in the service of his evangelical and ecumenical approach to race salvation.

The loose coalition of old and new Nazis, segregationists, repatriationists, nativists, and recrudescent race scientists Cox joined must be interpreted with caution. As with any political movement, members brought quite distinct concerns, and support for the entire *weltanschauung* cannot be assumed. It is unlikely that Cox favored the imposition of a full National Socialist government, but his vision of a war against the Jews for the salvation of the white race made him a hero to neo-Nazis around the globe.

Cox's career illustrates how racial ideas of the late nineteenth and early twentieth centuries could be kept alive in the postwar years by the energetic formation of coalitions and alliances directed toward legislative aims. His appealing use of the history of nations and peoples helped forge these coalitions and served as a valuable, if ultimately unsuccessful, supplement to the postwar use of IQ data to support segregation. He had little interest in new biological or psychological data on race; history had already spoken decisively on the issue. Cox hoped for the salvation of the white race through legislation based on his version of biologized history. For many of those he worked with during the postwar and civil rights years up until his death, race salvation could not be achieved through legislation; a more radical political transformation was required. Cox's broad appeal stemmed in part from his unrelenting hope that whites *and* blacks would peacefully come to their senses and separate forever.

Cox's many activities and the diversity of his alliances might appear to be a bricolage, an opportunistic and rather slapdash collection of venues and ideas. However, for Cox, as for Knox, race was everything. The political order, the legal regime of a geographic area—*everything* could be reduced to the underlying racial foundation. Viewed from Cox's eyes, the call for repatriation, the removal of whites from Africa, and the legal critique of *Brown* were one and the same problem, since they were all, in a very real sense, constituted by race. His postwar career was not a case of how racial ideas interacted with legal/political ideas; rather it is an illustration of how legal/political ideas were viewed as racial ideas.

Notes

1. Robert Knox, *Races of Men: A Fragment* (Philadelphia: Lea and Blanchard, 1850), 7. For biographical information on Knox, see John P. Jackson, Jr., and Nadine M. Weidman, *Race, Racism, and Science: Social Impact and Interaction* (Santa Barbara, CA: ABC-Clio, 2004).
2. Knox, *Races of Men*, 250–51. On Knox's anticolonial stance, see Evelleen Richards, "The 'Moral Anatomy of Robert Knox: The Interplay Between Biological and Social Thought in Victorian Scientific Naturalism," *Journal of the History of Biology* 22 (1989): 373–36.
3. Joel Williamson, *The Crucible of Race: Black-White Relations in the American South Since Emancipation* (New York: Oxford University Press, 1984).
4. Madison Grant, *The Passing of the Great Race; or the Racial Basis of European History*, 4th ed. (New York: Scribners, 1921), 77.
5. Lothrop Stoddard, *The Rising Tide of Color Against White World Supremacy* (New York: Scribners, 1921), 227.
6. George M. Fredrickson, *Racism: A Short History* (Princeton, NJ: Princeton University Press, 2002), 109.
7. Richard H. King, *Race and the Intellectuals, 1940–1970* (Baltimore, MD: Johns Hopkins University Press, 2004), 310.
8. Audrey Smedley, *Race in North America: Origin and Evolution of a Worldview*, 2nd ed. (Boulder, CO: Westview Press, 1998), 213. Even from the beginning of repatriation efforts in the nineteenth century, whites had difficulties persuading African Americans of the desirability of moving to Africa. See Jacqueline Bacon, "'Acting as Freemen': Rhetoric, Race, and Reform in the Debate Over Colonization in *Freedom's Journal*, 1827–1828," *Quarterly Journal of Speech* 93, no. 1 (2007): 58–83.
9. Williamson, *Crucible of Race*, 219 (see n. 3).
10. Biographical details are from Ethel Wolfskill Hedlin, "Earnest Cox and Colonialization: A White Racist's Response to Black Repatriation, 1923–1966" (PhD diss., Duke University, 1974) and Cox's privately published autobiography, "Black Belt Around the World" (Richmond, VA, 1963).

11. Cox, "Black Belt," 328–29 (see n. 10).
12. "Bantu Negroes Brightest," *New York Times*, December 15, 1915, 24.
13. W. E. B. DuBois, "Lies Agreed Upon," *Crisis* 11, no. 4 (February 1916), 187.
14. Earnest Sevier Cox, *White America* (Richmond, VA: White America Society, 1923). On the move to self-publish, see Hedlin, "Earnest Cox and Colonialization," 47 (see n. 10).
15. Jonathan P. Spiro, "Patrician Racist: The Evolution of Madison Grant" (PhD diss., University of California–Berkeley, 2000), 572.
16. Cox, *White America*, 23 (see n. 14).
17. Ibid., 26, 27.
18. Ibid., 231.
19. Jackson and Weidman, *Race, Racism, and Science*, 129–62 (see n. 1).
20. For a useful introduction to the context of Gobineau's ideas, see Ivan Hannaford, *Race: The History of an Idea in the West* (Baltimore, MD: Johns Hopkins University Press, 1996).
21. On Chamberlain's interest in history as social action, see Geoffrey G. Field, *Evangelist of Race: The Germanic Vision of Houston Stewart Chamberlain* (New York: Columbia University Press, 1981), 174.
22. William H. Tucker, *The Funding of Scientific Racism: Wickliffe Draper and the Pioneer Fund* (Urbana: University of Illnois Press, 2002), 15–16.
23. See Gregory Dorr, "Assuring America's Place in the Sun: Ivey Foreman Lewis and the Teaching of Eugenics at the University of Virginia, 1915–1953," Journal of Southern History 66 (2000), 257–96.
24. Major biographies of Garvey include David Cronon, *Black Moses: The Story of Marcus Garvey and the Universal Negro Improvement Association* (Madison: University of Wisconsin Press, 1969); Elton C. Fax, *Garvey: The Story of a Pioneer Black Nationalist* (New York: Dodd, Mead, 1972); and Judith Stein, *The World of Marcus Garvey: Race and Class in Modern Society* (Baton Rouge: Louisiana State University Press, 1986).
25. Histories of Cox's relationship with Garvey include Barbara Bair, "Remapping the Black/White Body: Sexuality, Nationalism, and Biracial Antimiscegenation Activism in 1920s Virginia," in Sex, Love, Race: Crossing Boundaries in North American History, ed. Martha E. Hodes, 399–422 (New York: New York University Press, 1999); and William A. Edwards, "Racial Purity in Black and White: The Case of Marcus Garvey and Earnest Cox," Journal of Ethnic Studies 15 (1987): 132–33.
26. Tucker, *Funding of Scientific Racism*, 33–39 (see n. 22).
27. For Bilbo's views on race, see Theodore Bilbo, *Take Your Choice: Separation or Mongrelization* (Poplarville, MS: Dream House, 1947), which features an introduction by Cox.
28. Michael W. Fitzgerald, "'We Have Found a Moses': Theodore Bilbo, Black Nationalism, and the Greater Liberia Bill of 1939," *Journal of Southern History* 63 (1997), 293–320. This entire section has been adapted from John P. Jackson, Jr., *Science for Segregation: Race, Law and the Case Against Brown v. Board of Education* (New York: New York University Press, 2005).
29. Reginald Kearney, *African American Views of the Japanese: Solidarity or Sedition?* (Albany: State University of New York Press, 1998), 97.

30. Ibid., 98–99.
31. "Subject: Peace Movement of Ethiopia," 4 June 1942, FBI Files on Peace Movement of Ethiopia, FOIPA no. 1039474-000.
32. "The Peace Movement of Ethiopia, Also Known as Ethiopian Peace Movement," 30 September 1942, FBI Files on Peace Movement of Ethiopia, FOIPA no. 1039474-000.
33. "Peace Movement of Ethiopia: Internal Security, Sedition," 16 September 1942, FBI Files on Peace Movement of Ethiopia, FOIPA no. 1039474-000.
34. Ernest Allen, Jr., "Satokata Takahashi and the Flowering of Black Messianic Nationalism," *Black Scholar* 24, no. 1 (1994): 23–46.
35. "Internal Security, Sedition" (see n. 34).
36. See, for example, "Internal Security, Sedition" (see n. 34).
37. "Seize 84 Negroes in Sedition Raids," *New York Times*, September 22, 1942, 22.
38. "Also Known as Ethiopian Peace Movement," 5 November 1942 (see n. 33).
39. Ibid.
40. Kearney, *African American Views of the Japanese*, 100, 201 (see n. 30).
41. Using his family fortune, Draper created the Pioneer Fund in 1937 to promote racial homogeneity in the United States. For the history of the Pioneer Fund, see Tucker, *Funding of Scientific Racism* (see n. 22); and Paul A. Lombardo, "'The American Breed': Nazi Eugenics and the Origins of the Pioneer Fund," *Albany Law Review* 65 (2002): 743–830.
42. See Hedlin, "Earnest Cox and Colonialization," 82 (see n. 10).
43. Johann von Leers to Earnest Sevier Cox, 23 March 1955, Earnest Sevier Cox Papers, box 10, Rare Book, Manuscript, and Special Collections Library, Duke University, Durham, NC.
44. Jackson, *Science for Segregation*, 40–41 (see n. 29).
45. Joseph W. Bendersky, *The "Jewish Threat": Antisemitic Politics of the U.S. Army* (New York: Basic Books, 2000); and Norman Cohn, *Warrant for Genocide: The Myth of the Jewish World Conspiracy and the Protocols of the Elders of Zion* (London: Serif, 1967).
46. Willis Carto to Earnest Sevier Cox, 29 March 1954, and Cox to Carto, 1 June 1954, Earnest Sevier Cox Papers, box 9, Rare Book, Manuscript, and Special Collections Library, Duke University, Durham, NC. On Carto, see Frank Mintz, *The Liberty Lobby and the American Right: Race, Conspiracy and Culture* (Westport, CT: Greenwood, 1985).
47. Earnest Sevier Cox to Archbishop Addison, 15 July 1962, Earnest Sevier Cox Papers, box 15, Rare Book, Manuscript, and Special Collections Library, Duke University, Durham, NC.
48. Earnest Sevier Cox to *White Sentinel*, 14 October 1962 (see n. 48). On Smith, see Glen Jeansonne, *Gerald L. K. Smith: Minister of Hate* (Baton Rouge: Louisiana State University Press, 1997).
49. Earnest Sevier Cox to Mrs. Lambert Schuyler, 19 March 1959, Earnest Sevier Cox Papers, box 26, Rare Book, Manuscript, and Special Collections Library, Duke University, Durham, NC.
50. *American Mercury*, originally founded by H. L. Mencken, passed to the control of the John Birch Society in the 1950s and came under Carto's control in the early 1970s.

51. For information on Roger Pearson and the Northern League, see Jackson, *Science for Segregation* (see n. 29); Tucker, *Funding of Scientific Racism* (see n. 22); and Andrew Winston, "Science in the Service of the Far Right: Henry E. Garrett, the IAAEE, and the Liberty Lobby," *Journal of Social Issues* 54 (1998): 179–210.
52. Earnest Sevier Cox to Johann von Leers, 7 June 1955 (see n. 44).
53. Roger Pearson to Earnest Sevier Cox, 8 February 1958, Earnest Sevier Cox papers, box 3, Rare Book, Manuscript, and Special Collections Library, Duke University, Durham, NC.
54. Earnest Sevier Cox to Alastair Harper, 6 April 1960, Earnest Sevier Cox papers, box 14, Rare Book, Manuscript, and Special Collections Library, Duke University, Durham, NC; and Cox, "Moot," 14 September 1959, Earnest Sevier Cox Papers, box 25, Rare Book, Manuscript, and Special Collections Library, Duke University, Durham, NC.
55. William Tucker, *The Science and Politics of Racial Research* (Urbana: University of Illinois Press, 1994); and Jackson, *Science for Segregation* (see n. 29).
56. Michael J. Klarman, "How *Brown* Changed Race Relations: The Backlash Thesis," *Journal of American History* 81, no. 1 (1994), 82. See also Klarman's more extended presentation of his argument in Klarman, "*Brown*, Racial Change, and the Civil Rights Movement," *Virginia Law Review* 80 (1994): 7–150; and Klarman, *From Jim Crow to Civil Rights: The Supreme Court and the Struggle for Racial Equality* (New York: Oxford University Press, 2004). Other standard sources on the political reaction to *Brown* are Numan V. Bartley, *The Rise of Massive Resistance and the Politics of the South During the 1950s* (Baton Rouge: Louisiana State University Press, 1969); and Francis M. Wilhoit, *The Politics of Massive Resistance* (New York: George Braziller, 1973).
57. David R. Goldfield, *Black, White, and Southern: Race Relations and Southern Culture, 1940 to the Present* (Baton Rouge: Louisiana State University Press, 1990), 79.
58. The best treatment of the Citizens' Councils remains Neil R. McMillen, *The Citizens' Council: Organized Resistance to the Second Reconstruction, 1954–64*, repr. ed. (1971; repr., Urbana: University of Illinois Press, 1994).
59. S. R. Booth to Earnest Sevier Cox, November 1955 (see n. 50).
60. Oak Cliff White Citizens' Council Library to "Dear Friend," n.d. [but probably 1958], Earnest Sevier Cox Papers, box 13, Rare Book, Manuscript, and Special Collections Library, Duke University, Durham, NC.
61. Earnest Sevier Cox, *Unending Hate: Supreme Court School Decision a Milestone in the Federal Program to Break the Will of the White South in Its Dedicated Purpose to Remain White* (Richmond, VA, 1955). According to Cox's correspondence, *Unending Hate* had been completed by January 1955, seven months after the *Brown* decision. See Earnest Sevier Cox to Willis Carto, 3 January 1955 (see n. 44).
62. Cox, *White America*, 357 (see n. 14).
63. Earnest Sevier Cox, *The South's Part in Mongrelizing the Nation* (Richmond, VA: White America Society, 1926), 31.
64. Hedlin, "Earnest Cox and Colonialization," 171–213 (see n. 10).
65. Earnest Sevier Cox to Wilhelm Ladewig, 18 November 1958 (see n. 61); and Earnest Sevier Cox to Willis Carto, 3 January 1955 (see n. 44).

66. Fitzgerald, "'We Have Found a Moses,'" 298 (see n. 29).
67. Earnest Sevier Cox to Willis Carto, 24 January 1955 (see n. 44).
68. Reginald Horsman, "Origins of Racial Ango-Saxonism in Great Britain Before 1850," *Journal of the History of Ideas* 37 (1976): 387–410.
69. Thomas R. Gossett, *Race: The History of an Idea in America* (Dallas, TX: Southern Methodist University Press, 1963), 84–122.
70. Cox, *Unending Hate*, 17 (see n. 62).
71. *Brown v. Board of Education* 347 U.S. 483 (1954).
72. Mark V. Tushnet, *Making Civil Rights Law* (New York: Oxford University Press, 1994), 314–15. On the origins of the specific studies cited and the collaboration between social scientists and attorneys in the *Brown* litigation, see John P. Jackson, Jr., *Social Scientists for Social Justice: Making the Case Against Segregation* (New York: New York University Press, 2001).
73. Cox, *Unending Hate*, 41 (see n. 62).
74. Ibid., 42.
75. Ibid., 41–42.
76. Earnest Sevier Cox to Willis Carto, 8 June 1964 (see n. 48).
77. Earnest Sevier Cox to Matt Koehl, 20 February 1964 (see n. 48).

Mongrels and Hybrids: The Problem of "Race" in the Botanical World

Vassiliki Betty Smocovitis

My difficulty probably is that I do not know the exact meaning of words like species, speciation, populations, reproductive isolation, allopatric and sympatric speciation, geographic and ecological isolation as applied in your paper. If I am wrong in my interpretation of them would you therefore kindly correct me. I must confess that I always have difficulties in reading papers discussing terminology, for I have in mind that the living world does not lend itself too well in our rigid classifications. I feel therefore that all kinds of transitions exist between sympatric and allopatric distribution as well as between different kinds of species. Scientific progress has often been delayed by a premature freezing of our concepts into a terminology that hinders our clear thinking.

—Jens Clausen, botanist, writing to Ernst Mayr, zoologist[1]

The problem of race hasn't figured prominently in the entire biological world. In the plant world, for example, it rarely figures at all. What does figure, especially since Darwin first noticed it, is the deeper problem of variation. What it is, its origin, its maintenance and preservation (or its heritability), its pattern (continuous or discontinuous), and ultimately what biological purpose or function it serves does matter greatly. Common sense would tell us that if organisms were in fact adapted to suit their environments, then we would expect to find perfectly adopted forms, not all kinds and manners of endless variability.

The fact of the matter is that the problem of variation has been at the heart of evolutionary thinking, yet few historians of evolution have directly studied it. We know, for example, that Darwin drew the distinction between variation under nature and variation under domestication in much the same way that he drew the distinction between natural and artificial selection.[2] We also know that he recognized its importance to his theory of descent with modification because he devoted an entire two volumes to the subject titled *Variation of Animals and Plants Under Domestication*. We also know that it was at the heart of the celebrated debate between his later intellectual supporters, the biometricians, and his detractors, the Mendelians, at the turn of the century, as well as the source of Hugo de Vries's popular mutation theory at the turn of the century.[3]

Simply understood, the problem of variation was a central concern for evolutionists after 1859.[4] Questions posed included the following: What was its origin? How was it maintained? What purpose did it all appear to serve? Why, if adaptation were the case, did not one perfect, ideal form exist and occupy each particular zone? That question, of course, undergirded much of Darwin's thinking while he explored the natural history of the world during his five-year voyage aboard the HMS *Beagle*. Answering it took more than twenty years and more than four hundred pages of text organized by the familiar title *On the Origin of Species by Means of Natural Selection*, with the less familiar subtitle *Or the Preservation of Favoured Races in the Struggle for Life*.

If Darwin's famous title is an accurate indication, *race* was intended to be a category that applied to all living organisms, which in the nineteenth century mostly included plants and animals. Throughout the book, Darwin drew generously on examples from both the animal and the plant world, yet strangely enough, he did not in any systematic sense apply the term *race* to plants and mentioned it only rarely in the case of animals. Like others before and after him, terms like *variety* and *subspecies* or *species* were used instead when reference was made to the stages in plant evolution and even more generally to the discernable stages of evolution. Chapter 2 of *Origin of Species*, titled "Variation under Nature," which outlines the stages in the origin of species, notes the grade from individual differences to lesser varieties to well-marked varieties to subspecies to species. Differences between these groups "blended" into each other in an "insensible" series. Darwin wrote, "No clear line of demarcation has as yet been drawn between species and sub-species . . . or, again between subspecies and well marked varieties, or between lesser varieties and individual differences."[5] Defining *species* proved to be difficult, however, and Darwin famously skirted any attempt at a rigorous definition.[6] He wrote, "It will be seen that I look at the term species, as one arbitrarily given for the sake of convenience to a set of individuals closely resembling each other, and that it does not essentially differ from the term variety, which is given to less distinct and more fluctuating forms."[7] (This is ironic indeed, given the book's title.)

It is a small wonder, then, that when Theodosius Dobzhansky turned to the subject in the late 1930s, he titled his book *Genetics and the Origin of Species*.[8] It was his intention to remedy the fact that Darwin didn't properly discuss the origin of species, nor did he actually attempt to provide a true definition. The problem of defining species remained difficult, of course, and indeed intractable until the 1930s. Darwin understood that it remained one of the weakest parts of his theory and noted that his understanding of species was problematic and that evolution from the plant world made that understanding even more difficult. To Darwin, plants had made splendid examples, but at times they made his project even more complicated because they appeared to demonstrate so many special cases and anomalies.[9] As an example, Darwin was acutely aware of the problem of hybridization and how it could challenge the tidy branching patterns he tried to discern for his picture of general evolution. This much was stated in a special chapter—chapter 8—he devoted specifically to the subject of hybrids, which drew mostly on examples from plants. For Darwin, *hybrid* was a term that denoted a cross between species, while the term *mongrel* was used to denote any cross between varieties. Darwin never properly resolved the issue of hybridization in his book, nor did many of his successors. At the turn of the century, for example, as knowledge of genetics drew attention to the origin of variation on a physiological level, botanists turned seriously to hybridization as explaining the origin of variation and, thenceforth, the origin of new species. Explored by individuals like J. P. Lotsy in the years 1910–1919 and 1920s or so in books like *Evolution by Means of Hybridization*, hybridization was thought to be the specific source of raw material that generated variation.[10] If selection acted, it was to cull out the less adapted forms. Lotsy wrote, "Crossing therefore is the cause of the origin of new types, heredity perpetuates them, selection is the cause—not of their origin as was formerly supposed—but of their extinction."[11] Because crossing and hybridization were central evolutionary processes, any attempt at creating a natural or phylogenetic method in taxonomy was therefore futile. Lotsy wrote, "Phylogeny e.g. reconstruction of what has happened in the past is no science but a product of phantastical speculations which can be held but little in check by the geological record, on account of the incompleteness of the latter."[12]

Lotsy himself is now regarded as a marginal and obscure figure, but at the turn of the century he was part of a large community of "planty" people (I hesitate to call them all "botanists" since many only used plants for utilitarian ends) who rejected strict notions of Darwinian evolution in favor of an alternative number of theories to account for the origin of species. Not only did plant evolutionists espouse evolution by means of hybridization and embrace theories like de Vriesian *Mutationstheorie*, but a large number also remained ardent neo-Lamarckians, accepting the direct modification of characters by the environment, well into the 1930s. Individuals like Frederic Clements, the celebrated ecologist, and others studying the variation patterns of plants in nature saw what appeared to be the direct modification of the environment on plant characters.

The Biological Attributes of Plants

What was it that made plant evolution so intractable? While they were the preferred study or model organism (to use a modern term) for a number of evolutionists and especially geneticists in the early years of the twentieth century, plants also displayed phenomena unusual and quite different from animals. According to contemporary biological understanding, there are at least six notable differences between plants and animals:[13]

1. Plants tend to be developmentally simpler than animals. Animal systems are generally considered more complex because they require close integration between the various organs and organ systems required for functions not apparent in animals (e.g., behavior, motility, sense perception, and the coordination and balance of such complex functions). Plant hybrids are therefore easier to attain because animal hybrids require great coordination between complex parts.

2. Plants cross-pollinate "promiscuously" (in botanical terms), making reproductive isolation more difficult. This is in contrast to animals, which are able to establish barriers to species interbreeding more readily because of phenomena such as behavior motility and sense perception. These phenomena provide opportunities for preserving species boundaries or creating new species through reproductive isolation.

3. Plants also have an open or indeterminate system of growth and development. This is because they bear meristematic tissues (a kind of embryonic stem cell at both the root and shoot and sometimes in axillary buds). Indeterminate development allows plants to amplify the number of body parts they have and their overall size, and it can make possible greater individual longevity as well as asexual or vegetative reproduction. In plants, it is possible to establish a large clonal population from one individual mutant or a hybrid since sexual reproduction is not required. The establishment of a clonal population is also enabled by the fact that in plant species, low sexual fecundity does not necessarily impair the establishment of a plant population. The same open system also provides the opportunity for genetic mosaics through random mutations in the meristematic tissues. Unlike animals, one individual plant could therefore potentially generate a population of genetically different descendants.

4. Plants demonstrate much greater phenotypic plasticity than animals do. Plants vary more widely and respond rapidly to

environmental shifts. (Good examples of this include the variation between sun and shade leaves on trees and the difference in the leaf shape of plants submerged and those not submerged in aqueous environments or exposed to air).

5. Though it is not unique or exclusive to the plant world, plants demonstrate the phenomenon known as polyploidy, or the doubling or multiplication of chromosome sets. Through this mechanism, it is possible to have a sterile hybrid give rise to fertile progeny, which may in turn serve as founding members of a new species.

6. Finally, many plant species are capable of self-fertilization. They may be considered biological "hermaphrodites," for which uniparental reproduction is the typical method of sexual reproduction.

For all these "biological" reasons, as we now understand them, gaining a good understanding of the basics of plant evolution was thus especially difficult for geneticists, ecologists, and taxonomists after Darwin. Determining what counted as species and discriminating between phenotypic and genotypic responses to the environment were problematic in plants; but even the general pattern of variation in evolution, which in plants could be reticulating or networklike rather than dendritic or treelike, because of hybridization or the phenomenon of introgression, posed a number of complications in deriving a general theory of evolution.[14] (See figure 1.) Yet the need to create such a general theory was acute, since plants were ideal study organisms for biologists. As researchers relied more and more on plants as study or model organisms, plants' peculiarities or anomalies in comparison to organisms like birds and mammals became apparent.

It isn't surprising, therefore, that it took nearly one hundred years after Darwin first addressed the problem of variation for a general theory of variation and evolution in plants to be addressed. This appeared in 1950, under the unsurprising title *Variation and Evolution in Plants.*[15] The book was written by George Ledyard Stebbins, Jr., on the occasion of the invitation to deliver the Jesup Lectures at Columbia at the request of mammalian geneticist L. C. Dunn and Dobzhansky, Stebbins's lifelong friend and colleague. The book appeared at the tail-end of the interval of time between approximately 1930 and 1950 that saw a number of texts written that contributed to the overall synthesis of evolution, which historians have designated as the "Evolutionary Synthesis," or the synthesis between Darwinian selection theory and Mendelian genetics, in the hope of explaining the origin of biological diversity.[16]

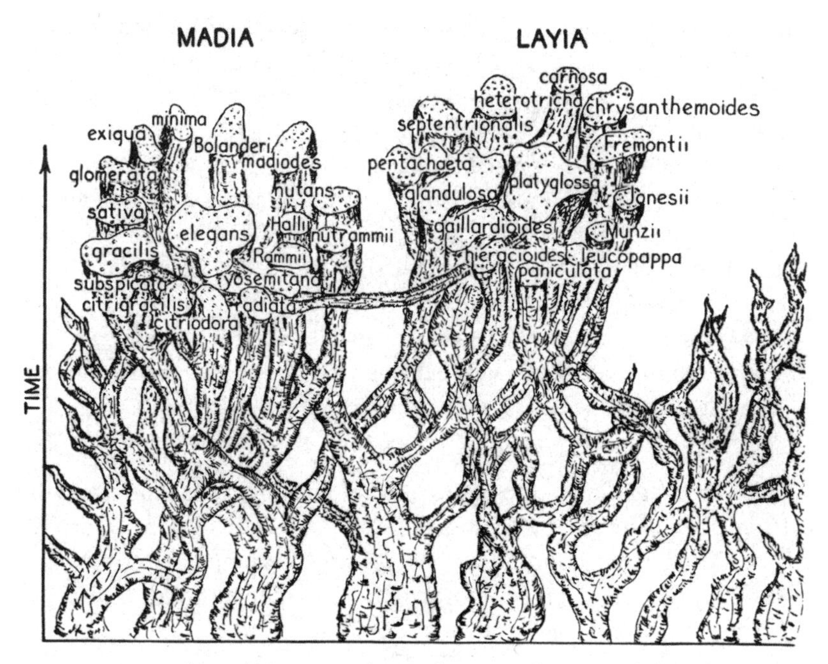

Figure 1. Hypothetical diagram of reticulating evolution and the relationships between the species of two genera of *Madia* and *Layia* at the present and projected back in time. (Jens Clausen, *Stages in the Evolution of Plant Species* [New York: Hafner Publishing, 1951], 179.)

Stebbins's book was the last in a number of such important texts, establishing what would be called "the synthetic theory of evolution" (which is sometimes referred to as neo-Darwinism or the new or modern synthesis of evolution). It was the only book on evolution that was taxon defined—meaning it was the only one organized around a specific biological group, namely plants. It was also the longest of all the books (at nearly 643 pages, which included nearly 1,250 references to a disparate body of literature on the subject). The first four chapters were organized around the general subject of variation patterns in plants, and the account of variation understood at the genetic level formed the remainder of the book. The subject of species and speciation was the subject of some three chapters, and the problem of defining a species was discussed explicitly a number of times in the book. The final set of chapters included a discussion of the big picture of evolution, namely the fossil history of plants with an eye to explaining macroevolution, in terms of the microevolutionary theory developed in earlier chapters.

Stebbins was closely following the work of Dobzhansky, the key architect of the evolutionary synthesis.[17] Stebbins's book in 1950 closely followed the arguments and organization Dobzhansky followed in his 1937 *Genetics and the Origin of Species*. It was in fact a kind of consistency argument—bringing into line (either by agreement or elimination) an understanding of plant evolution within a larger understanding of animal evolution. Dobzhansky and Stebbins had been friends and had been drawn together because of shared interests in formulating a general evolutionary theory and because their own organismic systems showed evolutionary compatibility. Dobzhansky's own research organism had been the famous *Drosophila pseudoobscura* while Stebbins had worked on a number of plant genera, including the genus *Crepis*. But while Stebbins followed Dobzhansky closely in a number of ways, he did not follow Dobzhansky explicitly on the importance of race as a meaningful category for understanding variation. Dobzhansky used race explicitly in reference to biological and especially geographic races in a number of organisms, including his *Drosophila*. Stebbins did not bring up the category in any formal way in his work. A quick glance at even the index to both volumes reveals Dobzhansky's singling it out prominently while Stebbins did not appear to single it out at all. Instead, in the chapter titled "Variation Patterns," the one chapter that would have demanded such a consideration, Stebbins wrote a familiar refrain for plant systematists:

> *The experience of most plant systematists, as well as their zoological colleagues, has been that the recognition of infraspecific units of several degrees or rank, such as subspecies, variety, subvariety, and form, produces more confusion than order. Units of one rank, termed subspecies by all zoologists and many contemporary botanists, are enough to express the great majority of the biologically significant infraspecific variation that can be comprehended by anyone not a specialist in the group.*[18]

Dobzhansky's book echoed his own research program into the geographic races of *Drosophila pseudoobscura*. Dobzhansky had selected this organism because its evolutionary history, as revealed by adaptations to local environments, could easily be determined by mapping of the giant salivary chromosome.[19] But Dobzhansky didn't stop there. He actively discussed and endorsed the existence of geographic and biological races in other insects and "lower" forms of life like snails, drawing on the work of Alfred Kinsey's studies of variation in the Cynipedae, or the gall wasps, as well as John Thomas Gulick and Henry Edward Crampton's work on geographic variation in *Partula*, a complex genus of snails. Examples of racial variation also abounded in the dominant areas of zoology, like that dealing with birds and mammals, but Dobzhansky skirted the issue of race existing in plants (as we would expect), and what examples of plants he did include mostly referred to the conventions botanists used to group plants into varieties or subspecies. Microbes were, for the most part, absent from examples he drew on, which is also as we would expect, since little was known about their taxonomy in the 1930s.

Where *race* did gain some usage was in ecology, or more precisely in genecology, the study of plants under a variety of environmental conditions that was popular in the early decades of the twentieth century. Race here was usually used to denote different groups that seemed to be correlated with climate, geography, or edaphic conditions. As noted by D. Briggs and S. M. Walters in their comprehensive, popular book *Plant Variation and Evolution*, one person who used the term *petites especes*, which translated into "local races," in a paper of 1901 was F. Ludwig.[20] Until Ludwig, the term *local race* had been used "rather loosely for plants from particular areas used for biometrical study or experiments." Ludwig, however, sought to make these entities of local races "real" by using biometrical evidence based on the number of floral parts in *Ranunculus ficaria*. In the early issues of the journal *Biometrika*, Briggs and Walters pointed out that the "reality" of local races had been an important topic for discussion, though the reality of races was challenged by an editorial in 1902 that disputed the sampling of polymorphisms, which were deemed spurious; questions could be raised about what counted as a "locale" by collectors. The use of the term *race* continued, however, but in that "loose" sense of the term and was usually associated with climatic, edaphic or geographic parameters.[21]

In the mid-1930s, the term *race* was used, though loosely again, by the celebrated interdisciplinary team of Jens Clausen, William Hiesey, and David Keck, though more than any other contemporaries, they challenged any strict definitions even of terms like *species*.[22] Plants seemed to have a more complex pattern of variation and evolution, and they defied any simple or rigid categorization that might work for birds or mammals. The opening epigraph, a private exchange between Clausen and avian systematist Ernst Mayr, is a nice demonstration of the differences between many botanists and zoologists in the middle decades of the twentieth century, though to be sure, some of the distinctions or differences were glossed over by the drive to create a coherent view of evolution. In one celebrated instance, Mayr, reviewing the latest monograph by the Carnegie Institution of Washington at Stanford team of Clausen, Hiesey, and Keck, showed how nicely birds, mammals, and even insects and plants could be integrated together, all recognizing the notion of race. He wrote:

> *It is evident from these studies that there is no fundamental difference between plants and animals in the evolution of populations and races that have become adapted to local environments. The difference is one of degree, with insects and plants having much more localized races than birds or mammals. That some of the insect races are as localized as the plant races is indicated by the work of Dobzhansky on Drosophila pseudoobscura. None of the insect material, however, can compare with some plant species in the abundance of conspicuous morphological characters in addition to the physiological ones.*[23]

Similarly, the contrast between Dobzhansky and Stebbins and their reliance on notions of race is telling. For most botanists, the word *race* held little meaning or utility, in part because the patterns of variability were more complex and the underlying mechanisms more complex and unlike those seen in animals. To be sure, some botanists had occasion to resort to the concept, but even then it was in a limited and questionable capacity.

The other instance where *race* was employed heavily was in designating the varied forms of an important crop plant: corn, or maize, as in *Zea mays*. Maize, a heavily modified (meaning "domesticated") plant, had a complex evolutionary history that could not in fact be easily separated from humans. Corn had been so heavily altered by human use that its own evolutionary history—and its taxonomy—remained obscure until the early to mid decades of the twentieth century. Its history was unraveled by methods involving both cytogenetics and anthropology (ethnography, to be more precise) as they came together in the area defining itself as ethnobotany. The efforts of maize historians like Paul Manglesdorf (probably the best known of the maize workers) and researchers like Edgar Anderson helped unravel corn's evolutionary history.

In 1942, in a paper published in the *Annals of the Missouri Botanical Garden* titled "Races of *Zea Mays*: Their Recognition and Classification," Anderson, along with Hugh Cutler, set out one of the first "natural" taxonomic schemes for corn that would reflect this evolutionary history. Up to that point, geneticists like E. Lewis Sturtevant had used an artificial system that had held to human patterns of use of corn types such as pop, flint, dent, flour, and sweet. To create what was a "true" and natural scheme based on corn's complex evolutionary history, Anderson and Cutler looked to the science of anthropology and the concept of race, substantively using the work of E. A. Hooton and his method of racial analysis, and one of its celebrated applications by Carleton Coon in his book *The Races of Europe*, published in 1939.[24] There were a number of parallels: both had been "domesticated" and both had a complex history. "The problem of races and their recognition," they noted, "is indeed almost the same in *Zea Mays* as in mankind. In both cases it is not easy to work out the racial composition of the whole, and it is difficult to give a precise definition of the term 'race.'"[25] Since Coon had discussed this "latter problem" in 1939, Anderson and Cutler directly referred to Coon in one long direct citation in their own paper.[26] The link to currents of thought in anthropology was not surprising; corn had been racialized by human practices, and drawing on some of the prevailing racial schemes like that of Coon's appeared to make sense for a plant whose racial qualities appeared obvious to anyone in the 1930s and 1940s wishing to create a classification scheme. Both were, after all, "domesticated," and old in their histories. Anderson and Cutler, therefore, adopted Coon's somewhat vague but popular definition of race to define the word "race as loosely as possible, and say that race is a group of related individuals with enough characteristics in common to permit their recognition as a group."[27] Despite its initial popularity, the term *race* and Coon's use of it began to gain some notoriety after 1963 when Coon published his book *Origin of the Races*. In it, Coon continued to draw on his initial formulation of race and suggested that five races of humans were in fact subspecies dating back to *Homo erectus*. Interestingly enough, Dobzhansky challenged Coon on his understanding of speciation and "race."[28] Anderson and Cutler's responses are unknown.

There is at least one other use of the term *race* in the plant world, and that is in mycology, the study of fungi, but this may not be all that surprising. The history of mycology has been dominated by the history of plant disease, and the history of plant disease is closely linked to agriculture and thence to human history. As in the case of corn, this is an instance where plants are either domesticated or seen through the lenses of human social practice. To the extent that it can be seen at all independently of humans, the biological world in general, and the botanical world in particular does not require the use of the category of race, and has in fact relied very little on it, especially given that plant taxonomy is a heavily worked field. Perhaps it is because of the staggering assortment of variation that exists in the plant world that makes all categories (and not just race) especially provisional, or perhaps because plants are so alien to humans, that categories like race, which are loaded with human meaning, are less easily transported; but there is little doubt that the use of *race* has had much less importance or meaning in the botanical sciences. To state this further, though variation is especially abundant and problematic in the plant world, plant evolutionary biologists have, for the most part, not relied on racial categories, preferring instead to speak of varieties, subspecies, cultivars, or "lines," and even then these terms are held provisionally (certainly more so than by some zoologists). The word *race* has been used but oftentimes in situations closest to human history.

If I were to try to conclude this paper with only one closing thought, it would be to echo the well-known sentiments of evolutionists and historians like Stephen J. Gould and to point out that race is not a biological category that has much meaning outside of human history.[29] To be sure, all categories are human inventions—attempts to order the world on our terms—but it is clear from the history of botany and my attempt here to understand the problem of race in plants—and its absence—that this is one explicitly nonbiological and cultural category that makes little sense apart from human or cultural history. I close with the thoughts of yet another forgotten plant taxonomist, the Harvard botanist Charles Weatherby. In a response to a questionnaire asking about the species problem in 1937, he spoke for many plant taxonomists who recognized the provisional nature of their work:

> *It looks to me as you were trying to generalize on the assumption that there is a basic uniformity in taxonomic groups. There is nothing of the sort. Taxonomy is only a glorified guess—an attempt to construct a cross-section of lines of descent in a form intelligible to the human mind. It always contains two variable quantities—the plasticity of animate nature and the differing points of view of the people who work at it. You can generalize successfully, if at all, only by keeping these facts constantly in mind. I suspect that the situation is best expressed by the old aphorism; the only general rule is that there is no general rule.*
>
> *Therein lies the fascination of taxonomy for those who like it. It is not a matter of mechanically applying a universal set of categories to given groups of facts. Each group has to work out anew the method by which he may best achieve that transforming of confusion into order which is the greatest satisfaction of pure taxonomy.*[30]

Notes

1. Jens Clausen to Ernst Mayr, 22 January 1948, Carnegie Team Papers, "Ecotype Concept," Missouri Botanical Garden Archives, St. Louis.
2. I am relying on the first edition of Darwin's *On the Origin of Species*. See Charles Darwin, *On the Origin of Species*, facsimile of the 1st ed. (John Murray, London, 1859; facs., Cambridge, MA: Harvard University Press, 1964).
3. William B. Provine, *The Origins of Theoretical Population Genetics* (Chicago: University of Chicago Press, 1971); Peter J. Bowler, *The Eclipse of Darwinism: Anti-Darwinian Evolution Theories in the Decades around 1900* (Baltimore, MD: Johns Hopkins University Press, 1983); Jean Gayon, *Darwinism's Struggle for Survival: Heredity and the Hypothesis of Natural Selection* (Cambridge: Cambridge University Press, 1998); and Bowler, *Evolution: The History of an Idea*, 3rd ed. (Berkeley: University of California Press, 2003).
4. For one prominent example of the importance of variation, see William Bateson, *Materials for the Study of Variation: Treated with Especial Regard to Discontinuity and the Origin of Species*, rev. ed. (London: Macmillan, 1894; repr., Baltimore, MD: John Hopkins University Press, 1992).
5. Darwin, *On the Origin of Species*, 51 (see n. 2).
6. Darwin may have also recognized that embedding the concept of "species" in an unfolding historical process like descent with modification (or what came to known as evolution) may have also freed the concept from the confines of a strict definition.
7. Darwin, *On the Origin of Species*, 52 (see n. 2).
8. Theodosius Dobzhansky, *Genetics and the Origin of Species* (New York: Columbia University Press, 1937).
9. For a more comprehensive discussion of Darwin's botany in his *Origin of Species*, see Vassiliki Betty Smocovitis, "Darwin's Botany in Origin," in *Cambridge Companion to Darwin's On the Origin of Species*, ed. Michael Ruse and Robert J. Richards, 216–36 (Cambridge: Cambridge University Press, 2008). For a historical account of plants and the Darwin family, see Peter Ayres, *The Aliveness of Plants: The Darwins and the Dawn of Plant Science* (London: Pickering and Chatoo, 2008). See also David Kohn, *Darwin's Garden: An Evolutionary Adventure* (New York: New York Botanical Garden Press, 2008).
10. J. P. Lotsy, *Evolution by Means of Hybridization* (The Hague: Martinus Nijhoff, 1916).
11. Ibid., 134.
12. Ibid., 140.
13. This list is compiled from a variety of standard sources in plant evolutionary biology. See the list of books consulted at the end of these notes.
14. Jens Clausen, *Stages in the Evolution of Plant Species* (Ithaca, NY: Cornell University Press, 1951).
15. George Ledyard Stebbins, Jr., *Variation and Evolution in Plants* (New York: Columbia University Press, 1950).
16. Vassiliki Betty Smocovitis, "Keeping Up with Dobzhansky: G. L. Stebbins, Plant Evolution and the Evolutionary Synthesis," *History and Philosophy of the Life Sciences* 28 (2006): 11–50.
17. Ibid.

18. Stebbins, *Variation and Evolution*, 33 (see n. 15).
19. I include the complete bibliography available on Dobzhansky in Smocovitis, "Keeping Up with Dobzhansky" (see n. 16).
20. See D. Briggs and S. M. Walters, *Plant Variation and Evolution*, 3rd ed. (Cambridge: Cambridge University Press, 1997); and F. Ludwig, "Variationsstatistiche Probleme und Materialen," *Biometrika* 1 (1901), 11–29.
21. These categories are still maintained in Verne Grant, *Plant Speciation*, 2nd ed. (New York: Columbia University Press, 1981). This is the only contemporary book to refer to "race" as a meaningful category. The word is used to describe variation in geographic races. In chapter three, titled "Geographical Races," Grant wrote, "Local variation grades into geographical variation. The distinction between the two is arbitrary but practical. Local races are the focus of population biological studies, whereas the broader racial groupings are of interest in systematics, and are recognized as the subspecies or sometimes the variety of formal taxonomy." In other words, "race" is used in a largely arbitrary and vague fashion.
22. For a history of the "Carnegie team" of Clausen, Hiesey, and Keck see Patricia Craig, *Centennial History of the Carnegie Institution of Washington* (Cambridge: Cambridge University Press, 2005).
23. Ernst Mayr, "Climatic Races in Plants and Animals," *Evolution* 2 (1948): 375–376.
24. E. A. Hooton, "Methods of Racial Analysis," *Science* 63: 75–81; and Carleton Coon, *The Races of Europe* (New York: Macmillan, 1939).
25. Edgar Anderson and Hugh Cutler, "Races of Zea Mays: 1. Their Recognition and Classification," *Annals of the Missouri Botanical Garden* 29 (1942): 69–88.
26. Anderson and Cutler cited Coon's statement from page 5 of *The Races of Europe*: "Since man is the oldest domestic animal . . . any attempt to classify him by a rigid scheme is immensely difficult and the scheme must be elastic if it is to work at all. Hence the term 'race' must also be elastic. We may recognize if we like, certain major races of the Old World such as the [Khoi-San sic] Bushman-Hottentot, the Pygmy, the Australoid, the Negro, the Mongoloid, and the White. Within each of these major racial groups there are, or have been, smaller entities, which may deserve the designation of race in a lesser sense. These smaller entities consist, for the most part of groups of people reasonably isolated, and developing into local physical enclaves . . . At what border point such an entity becomes a major race it is not always possible to say." Ibid., 71.
27. Ibid., 71.
28. For a historical analysis of this controversy, see John P. Jackson, Jr., "In Ways Unacademical: The Reception of Carleton S. Coon's *The Origin of Races*," *Journal of the History of Biology* 34 (2001): 247–85.
29. For one example of his views, see Stephen J. Gould, *The Mismeasure of Man*, rev. ed. (New York: W. W. Norton, 1996).
30. Edgar Anderson to Charles Weatherby, 28 August 1937, Anderson Papers, Missouri Botanic Garden Archives, St. Louis.

Additional Contemporary References Consulted

Arnold, Michael. *Natural Hybridization and Evolution* (Oxford: Oxford University Press, 1997).

Endler, John A. *Geographic Variation, Speciation and Clines* (Princeton, NJ: Princeton University Press, 1977).

Grant, Verne. *Plant Speciation*. 2nd ed. (New York: Columbia University Press, 1981).

Niklas, Karl J. *The Evolutionary Biology of Plants* (Chicago: University of Chicago Press, 1997).

The Roman Campaign of '53 to '55: The Dunn Family among a Jewish Community[1]

Melinda Gormley

"If there's one thing that's bound to affect the biological constitution of a people, it's the mating pattern."

—L. C. Dunn, 1960[2]

In August 1953 L. C. Dunn set off for Italy with his wife, Louise, and their son Stephen and in search of a small isolated community that would become the subject of genetic and anthropological research.[3] The family's central question was, "Does the custom of marriage within one's own faith have biological consequences?"[4] By October the Dunns had chosen a Jewish ghetto community in the center of Rome. They lived among the community members and examined them as research subjects for roughly nine months.[5] Dunn, a geneticist, had a sabbatical leave and chose his destination with the intention of spending a productive year doing fieldwork of collecting blood, urine, and saliva samples from human beings to analyze human evolution through population genetics. Stephen was a graduate student of social anthropology at Columbia University and would be conducting dissertation research by analyzing the community's history, economy, and social structure. Louise participated in their research by acting as "diplomatic head-of-mission and commissary," gathering demographic information such as surname, residence, and occupation from members of the Roman Jewish community.[6]

Several aspects of Dunn's life converged in Italy while he and his family lived among the Jewish community. He had gone there as a result of international racial misconceptions that stemmed from the American eugenics movement, Nazi racial hygienics, and scientific racism. What he got out of his time in Rome was data demonstrating the existence of an isolated, biologically unique Jewish community living in the midst of a larger, predominantly Catholic population. Based on his serological data, Dunn had proof of a point he had been arguing in his anti–scientific racism publications: cultural factors affect human evolution. Combining their biological and anthropological findings, this father-and-son pair summarized their conclusion best:

> *We think we have established that the Jews of the Roman ghetto did maintain their cultural and biological identity as a distinct subcommunity within a larger one, and that social forces can shape a man's biology.*[7]

The following is a history of an early attempt at an interdisciplinary field study of human evolution by a family whose personal and professional lives were intimately intertwined. Dunn undoubtedly performed a significant amount of work on this topic because he was interested in human genetics; however, his son's personal needs and professional career greatly influenced Dunn. He and his wife were simply concerned parents looking out for their son.

The Dunn Family

Dunn was a revered mouse geneticist studying development and an effective administrator overseeing various scholarly agencies. After receiving his doctorate in genetics from Harvard University in 1920, he went to Storrs Agricultural Experiment Station in Connecticut, where he performed most of his investigations on poultry. Dunn remained at Storrs until 1928, when he began a lengthy career in Columbia University's Zoology Department before retiring in 1962. He helped establish the Genetics Society of America and was its first president in 1932. In 1961 he was president of the American Society of Human Genetics. He was managing editor of *Genetics* from 1935 to 1940 and the *American Naturalist* from 1951 to 1960.

Dunn's commitment to science permeated beyond his career and discipline. He frequently used his scientific expertise and personal networks as a basis for his socio-political activism. He was a socialist and humanitarian, fighting for civil liberties and against social injustices. In terms of his activism, Dunn's efforts in helping European political refugees relocate to the United States provided him with an intimate knowledge of the consequences wrought by the Nazi Party's Nuremberg Laws.[8] He was an executive member of the Emergency Committee in Aid of Displaced Foreign Scholars for its entire duration from 1933 to 1945 and secretary of a similar organization, the Faculty Fellowship Fund, at Columbia University, which raised money from its local community and placed refugees into its faculty.[9]

Dunn recognized that Nazi propaganda about a "pure" Aryan race had left a lasting impression internationally and that the United States' powerful eugenics movement and other forms of scientific discrimination had convinced most people that science supported racial inequality. Dunn believed that "the blood myth dies hard, and still survives, almost by linguistic use."[10] By that he wished to convey that believing in inheritance through blood, rather than genes, was an incorrect but enduring notion. It led to misconceptions, such as the notion that pure races exist, which in turn gave support to the promotion of racial superiority and condemnation of miscegenation.[11]

Concerned about these commonly held falsehoods, Dunn felt compelled to educate others about the limitations of science and the boundary between scientific evidence and scientists' interpretations of data. He wrote several publications between 1946 and 1952, in which he scientifically explained heredity to a nontechnical audience while simultaneously undermining scientific racism and racial prejudices. He cowrote the first and most well-known of these publications, *Heredity, Race and Society*, with *Drosophila* geneticist Theodosius Dobzhansky, his close friend and colleague in the Zoology Department at Columbia. This book brought Dunn to the attention of the United Nations Educational, Scientific, and Cultural Organization (UNESCO), whose members sought authors for its series the Race Question in Modern Society. Dunn cultivated a strong partnership with UNESCO, writing an educational pamphlet titled *Race and Biology* and serving as rapporteur of the 1951 committee that revised UNESCO's original "Statement on Race."[12]

Dunn's publications on race, whether he wrote them himself or with Dobzhansky, provided a broad definition of race and retained race as a biological category. In *Heredity, Race and Society*, Dunn and Dobzhansky wrote that "races can be defined as populations which differ in the frequencies of some gene or genes."[13] In *Race and Biology*, Dunn defined a race as "a group of related intermarrying individuals, a population, which differs from other populations in the relative commonness of certain hereditary traits."[14] Dunn and Dobzhansky both thought that racial categories were arbitrary, stating that it was impossible to create static racial groups, and thus promoted a flexible view of human racial categories. "'Race' is not a fixed or static category," Dunn told his UNESCO readers, "but a *dynamic* one."[15] What mattered was a comparison of gene frequencies for a particular trait and an appreciation for the instability of traits.

What differentiated human from animal populations, according to Dunn, was that humans were subjected to the effects of not only biological forces but also social ones. Some of the cultural barriers keeping humans from intermarriage mentioned by Dunn and Dobzhansky foreshadowed Dunn's research in Rome: "Civilized as well as primitive men have certain rules which govern the choice of mates and these rules frequently prescribe marital unions within a specified group—a clan, a tribe, nation, religious denomination, economic or social circle."[16] Five factors that formed races, according to Dunn, were mutation, selection, adaptation, migration, and isolation. Only the last, specifically geographical or social isolation, influenced marriage patterns in the present day and thus was "the great race-maker" among humans.[17]

Dunn and Dobzhansky's concern about racist misuses of science and their interest in scientific investigations of human evolution and population genetics inspired them to start the Institute for the Study of Human Variation at Columbia. This interdisciplinary institute brought together specialists trained in social, biological, medical, and mathematical sciences for collaborative studies of human evolution. Genetics stood at the heart of their cross-disciplinary scheme, uniting plants, animals, and humans by using the modern evolutionary synthesis as the main conceptual pillar. Dunn directed the institute throughout its duration from 1951 to 1958, and under the institute's auspices, Dunn and his family conducted their field research in Italy.

Leslie Clarence Dunn and Louise Porter married in 1918 and had two sons, Robert (Bob) Leslie, born in 1921, and Stephen Porter, born in 1928. They raised their children in Connecticut and New York and usually spent part of the summer in Maine. The war years greatly affected many people's lives, and the Dunns were no different. When the United States entered World War II, Bob was old enough to enlist and did. He trained as a pilot for advanced aerial gunnery, and in March 1944 his squadron was deployed to bomb Germany. Dunn worried about his elder son's safety, and when Bob received furlough after completing fifty flights, Dunn put his commitments on hold for two weeks to go to Maine and enjoy his son's visit.[18] Stephen, who suffered from a physical handicap, went through his adolescence during the war years. Attending to his special needs took a lot of devotion from Dunn and Louise. The 1940s mark a turning point for the Dunn family. Bob asserted his independency, whereas Stephen and his parents' lives became more intertwined in these years before their Roman adventure.

In the 1940s Dunn attended to his usual scientific and professional tasks and increased his extracurricular campaigns, even though he suffered a heart attack in the summer of 1940 at the age of forty-six. Specific information about this medical episode and his other health problems is scant except for frequent and brief comments about sicknesses and operations. Sociologist Robert S. Lynd, Dunn's colleague at Columbia, worried about Dunn not only because of his ill health but also because of his tendency to take on too much. Lynd hoped that the university's provost would consider reassigning the chairmanship of the Zoology Department to someone else during the 1940–41 academic school year.

> *He [Dunn] is high-mettled and hard-working, and it would probably be hard to persuade him to relinquish the departmental chairmanship. But just because he takes on responsibility so seriously and works so hard, some of the rest of us ought to try to re-structure his official situation so that he can't go on and kill himself.*[19]

Dunn confirmed Lynd's premonition by serving as the Zoology Department's chairman from 1940 to 1946, stepping down to take a long-overdue and much-needed sabbatical leave.[20] Dunn's overexertion during the war years eventually caught up with him. When he left on this sabbatical, he was exhausted, and after recovering for more than six months in Arizona, he still felt a lack of energy.[21] When Dunn left New York for the desert in the early summer of 1946, he had intended to travel to Uppsala, Sweden, by the fall. Due to poor health, he stayed in Arizona longer than planned before traveling to Scandinavia for a six-month stay.

In Sweden, Dunn visited Gunnar Dahlberg at the State Institute of Human Genetics at the University of Uppsala.[22] In 1942 Dahlberg had published *Race, Reason and Rubbish*, which used science to attack the Nazi Party's support of anti-Semitism and a pure Aryan race. Dunn and Dobzhansky had just written *Heredity, Race and Society*, which also explained genetics in an attempt to undermine racism. In comparison to Dahlberg's book, Dunn and Dobzhasky's was more inclusive, analyzing scientific racism in general; the authors did not limit themselves to the aspects arising from Nazi propaganda. Uppsala revived and energized Dunn, who returned to New York ready to tackle new scientific problems. Shortly after his homecoming, he and Dobzhansky decided to add human evolution to their studies into mouse and *Drosophila* genetics and began forming what would become the Institute for the Study of Human Variation.[23] Armed with new material on human genetics, Dunn started teaching a course on human heredity that was geared mainly to premedical students. At this time, it was highly irregular for American medical students to study human genetics.[24]

Louise and Stephen accompanied Dunn to Arizona and Sweden. As a result, Dunn and Stephen spent lots of time together. Stephen had a severe form of cerebral palsy that affected his ability to perform daily physical tasks, and it would require him to use a wheelchair throughout his seventy-one years of life.[25] Stephen was taking courses in English, anthropology, and sociology, and Dunn would read in the library while waiting to help Stephen get from one class to another. During their time together, Dunn learned a lot about people who suffered from various disabilities.[26]

Louise gave birth to Stephen on March 24, 1928, and although she and Dunn knew by his first birthday that something was seriously wrong, doctors had not yet provided an accurate diagnosis.

> *This has been a most disorganized and unhappy month—Stephen and Bob have both been ill—we took Stephen to the Babies Hospital for observation but have no diagnosis yet after ten days—altho' he only stayed two days & seems better now. Whether it's only a "normal" retardation in development or has some specific nervous disorder back of it we don't know yet & it's been most trying for Louise.*[27]

Dunn and Louise visited various hospitals, clinics, and schools in the northeastern United States for answers. Stephen stayed at some of these places for extended periods of time. When he was two-and-a-half years old, his parents took him to a clinic in Boston, where he lived for two months while learning how to walk. When they placed Stephen at a live-in school in Ascutney, Vermont, so that he would "be with other children and with sympathetic teachers trained to take care of him," Dunn recognized that it was harder on Louise and him than on Stephen. Leaving Stephen for so long saddened Louise. She hated to do it but also thought it was best for her son.[28]

By the age of five Stephen underwent the first of many surgeries aimed at providing him with better muscle control, but this particular surgery failed, as would many of the future ones that he endured.[29] Some of Stephen's operations decreased his ability to perform everyday tasks. While Stephen recovered from an operation on his hands in 1943, Dunn would feed him lunch and Louise would attend to him at dinner. It took Stephen a long time to recover from this operation, as noted by Dunn three months later: "Recovery is slow business & Louise and I are simply extending the summer programs of keeping time clear for him—taking turns with exercises care & school attendance."[30]

Stephen required constant aid from one or two people well beyond his formative years, and although Dunn and Louise usually had someone to assist with him, they had a hard time finding caretakers during the war years. As a result, they took on a large portion of these daily tasks, such as feeding him and exercising his muscles. Stephen, a teenager during the war, started school near the Columbia campus in 1941, where Dunn could easily help him. After the war ended and the family had returned from Uppsala, Stephen started college at Columbia and eventually completed a doctorate in anthropology.[31] Years later, Dunn described his daily routines during Stephen's high school and college years.

> *I recall now that I was in & out of his school (3 blocks from my lab at Columbia) nearly every day & when he entered college[.] I at first brought him to Columbia each day then when he went to live in a dormitory for most of his course I had a habit of lurking in the background to be sure he got to his classes; & he formed the habit of dropping into my office for study & talk.*[32]

Dunn made a conscious decision to stay close to home during the war years. Stephen and Louise needed his help, and therefore he restricted his travel to short trips and local destinations.[33] Remaining near New York during Stephen's adolescence undoubtedly contributed to their strong bond and common interests that continued throughout their lives.

Dunn's exposure to Stephen's physical disability contributed to his growing disdain for eugenic policies in the late 1920s and early 1930s. Indeed, in 1929 Dunn was the only person skeptical of eugenics on a seven-member advisory committee for the Carnegie Institute of Washington, which evaluated the merit of data collected by the Eugenics Record Office. He was also the only member of the 1929 committee asked back in 1935 to reappraise the Eugenics Record Office's activities. Dunn supported the study of human heredity but warned against the premature and biased application of genetic theories.[34] He grew more vocal in his anti-eugenic statements as the 1930s progressed, and turned his attention to the general problem of scientific racism by the 1940s.

Although Stephen had extreme physical handicaps, he had a sharp mind and learned several languages, read extensively, and wrote poetry. Stephen knew five languages by the time he earned his undergraduate degree. He first learned the languages that Dunn knew: English, French, and German. When the Dunn family went to Oslo for Dunn's 1934–35 sabbatical leave, Stephen learned Norwegian. In the 1940s, Stephen started learning Russian and later translated Russian texts into English.[35] Both Dunn and Stephen tended to use written Italian for proper names of agencies that they worked with in Italy, but it is hard to say how much of the language each man learned while living there from 1953 to 1954.

The Soviet Union was another common interest father and son shared. Since Stephen attended school near Dunn's office in the 1940s, they met regularly throughout the day. Dunn and Dobzhansky, a Russian émigré, developed a close friendship beginning in the 1930s, and in 1940 Dobzhansky began his tenure at Columbia.[36] Not only was Dobzhansky one of Dunn's best friends, but they also worked closely together in aiding Soviet geneticists and performing scientific research. Stephen started learning to speak and read Russian in 1943. During the same year, Dunn actively worked toward improving communication with scientists in the Soviet Union by starting the American-Soviet Science Society, an agency that he presided over until 1946. One service provided by the society was translating Russian texts into English. For twenty-five years, beginning in the 1960s, Stephen founded and edited *Soviet Anthropology and Archeology*.[37] The majority of Stephen's research after he received his doctorate focused on Soviet society.

Poetry and literature also captivated father and son. Dunn had seriously considered a career in literature instead of zoology but was convinced to pursue biology, reasoning that a scientist can enjoy literature whereas a writer cannot perform rigorous scientific research. Starting in the 1930s, the family produced books of poetry together, published under the name of Coalbin Press. Dunn had acquired a printing press in 1904, and from the age of eleven through his college years, he had a small printing business. Eventually he gave the press to his two sons. Stephen had a sincere love of verse, and after returning from Italy, he printed five hundred copies of his poetry book, *Some Watercolors from Venice* in 1956. It was the fourth book published by Coalbin Press.[38]

After returning from Italy and before successfully defending his dissertation in 1959, Stephen experienced more setbacks and some good fortune. He underwent another one of his many surgeries in hopes of improving his physical handicaps. This particular surgery was an attempt to straighten his right wrist so that he might use crutches, but like so many other surgeries, it did not work.[39] Stephen married Ethel Deikman on October 6, 1956, after an eleven-year love affair.[40] Ethel also suffered from cerebral palsy, but she was not as severely impaired as Stephen. She could walk. They recognized their marriage as remarkable because people with disabilities rarely married, much less married someone with a handicap. They referred to their wedding as the Great October Revolution in acknowledgment of their personal fortune and mutual love of the Soviet Union.

The couple's professional interests overlapped to a great extent. They jointly published several books, including *The Peasants of Central Russia* in 1967 and *Introduction to Soviet Ethnography* in 1974. Stephen also wrote books independently that discussed the Soviet Union, such as *Cultural Processes in the Baltic Area under Soviet Rule* in 1966. Stephen and Ethel translated Russian publications into English. In addition to founding and editing *Soviet Anthropology and Archeology*, Stephen also translated Russian documents as editor of *Soviet Sociology*. Ethel noted that Stephen's favorite course to teach was Comparative Religions; however, he was rarely offered the chance to teach.[41]

Louise supported her family resolutely by attending to their needs and working with Dunn on some of his campaigns. For example, Dunn and Louise were joint treasurers of their local Russian War Relief fund in the early 1940s.[42] She also accompanied Dunn to foreign countries during all his sabbatical leaves, taking Stephen with them. Dunn remarked most frequently of Louise's devotion to Stephen. She extended her usual familial role while in Italy. She not only supported her husband and son while they worked, but also actively participated in data collection important to the scientific outcome. The Dunn family often traveled together, and as a result of Stephen's special needs, father, mother, and son were extremely close. Heading to a foreign country as a family was not new for these three; however, their activities in Italy added a new dimension to their family dynamics.

The Jewish Ghetto Community of Rome

Dunn and Stephen enjoyed a very close father-son relationship, so it is no surprise that Dunn proposed a joint study to his son by early 1952. Stephen was not interested initially but eventually changed his mind. He would not have been able accomplish field research without his father and mother's help.[43] Years later in the acknowledgments of his dissertation, Stephen recognized his parents' role in helping him graduate: "Last, but not least, my parents provided me with the opportunity to do the field-work in the first place. Whatever value this monograph may have, they are quite largely responsible for."[44] Similarly, Dunn and Louise felt blessed to have Stephen in their lives because he gave them "a great deal of both mental & physical stimulation & much happiness." Stephen's spirit and motivation greatly inspired his parents.[45]

Having a good idea of the kind of research project he wished to undertake, Dunn focused on choosing the physical location and research subject. The project would analyze a nuclear community, which he considered a manageable research project because of such a community's small mating circles and tendency to exhibit inbred disorders. Marriages seldom occur between members of a small, isolated community and members of the surrounding community, and Dunn found this particularly interesting because he considered this true only of *Homo sapiens*. Thus, Dunn consciously sought out an isolated human community because he believed that isolation, whether geographical or cultural, provided the best historical explanation for the development of human races. As mentioned earlier, Dunn explained in his race publications that human races are a product of five evolutionary processes—mutation, selection, adaptation, migration, and isolation—and that isolation provided the best explanation for the development of human races. Moreover, Dunn pinpointed several isolating factors at work among human populations: proximity, language, religion, education, and class or caste.[46]

Civilization, a recent historical development, brought humans into greater contact with one another and broke down geographical barriers. As a result, civilization increased racial intermixing and decreased clear racial distinctions. Geographical isolation of distinct biological races was becoming rarer as time passed; however, cultural factors implemented by choice or by force continued to segregate communities living in close proximity. Racial segregation of peoples with black and white skin in the United States and caste segregation in India were examples of forced cultural isolation that shaped mating patterns.[47]

It was the role of religion as a cultural factor influencing marriage decisions that intrigued Dunn and led to the family's analysis of the Jewish community of Rome. Although he had supported this idea in his writings on race, he had not yet investigated it himself. He, Stephen, and Louise would attempt to learn more about the influence that culture had on human genotypes, and thus on biological evolution, by studying a Jewish community near the Vatican in Rome. Two communities living in the same urban setting yet virtually isolated from each other because of their different religions provided Dunn with a manageable project for testing his theory.

Several factors contributed to the Dunns' decision to go to Italy and study a Jewish enclave. For one, Italy had more isolated communities than many other modern countries, and two, Dunn wanted to go to Italy. A Jewish population made the most sense because of its reliable historical records and the lengthy duration of Jewish traditional culture.[48] Reflecting on the family's choice of the Roman Jewish community, Dunn noted that the family did not choose this particular ghetto community because its members were Jewish. Rather, there was sufficient historical data on their isolation, and the other communities on which the Dunns had gathered preliminary background information had a greater degree of out-marriage than Dunn desired.[49]

Stephen was convinced early on that he and his father would analyze a Jewish community; furthermore, Stephen ascertained that the Roman Jewish community was their best choice based on comments made by a Dr. Anav. A year before leaving New York, the Dunns sent letters to various community leaders asking about marriage patterns of their community members. In a letter to Dunn, Dr. Anav, secretary of the Roman Jewish community, wrote that he had no objections to the study but also noted that the ghetto enclave did not satisfy the study's standards. According to Dunn, Dr. Anav wrote "in very courtly Italian" that "he wanted to draw my attention to the fact that it [the community] didn't meet the criteria, since they had been subjected to many baptisms, mixed marriages, and similar disgraces." It was Stephen who believed that Dr. Anav's last words were quite telling, and according to Dunn, Stephen made a remark to the effect of "That's where you'll find the community. If they maintain this spirit then that's the place. There might be some integrity in the community, and that's where we should go."[50] Stephen's hunch was correct. Dunn's analysis would determine that less than 5 percent of the community's members had descended from out-marriage.[51]

Getting to Italy, however, was not easy. Ruth Shipley of the Department of State tentatively denied Dunn's application for a passport in April 1953 because of his alleged Communist activities. Shipley's letter did not surprise Dunn,[52] and he replied promptly. Not only did he mention having previously stated under oath that he was not and never had been a Communist, but he also validated his past actions by summarizing his opinions and intentions in the following manner:

> *(1) a lifelong belief that civil liberties lie at the very foundation of democratic government; (2) an aversion and active opposition to Fascism which began when Hitler, Mussolini and later Franco destroyed civil liberties and attempted to destroy democratic government; (3) an active belief in and practice of international cooperation in science which extends over my whole scientific life.*[53]

Dunn received his passport five weeks later, and their original plans were not delayed.[54]

Due to his physical disabilities, the trip was onerous for Stephen, but the family made light of it. In Italy they had access to only cold water, which proved problematic to Stephen's physical condition and provided the rest of the family with an invigorating experience.

> *The achievement of a real steamy satisfaction is the event for which we all wait. The chief sufferer is Stephen. We others clamber into that icy tub (privatim et seriatim et maledictim) turn on the hand spray, and by dint of scrubbing and scraping and making loud noises we get the dirt off and get our exercise to boot. But this is no good for Stephen, whose muscles contract in cold water so that he can't sit up or stretch out and is only miserable.*

Nonetheless, the three of them, at Louise's instigation, fabricated an ideal solution: the Caracalla Cleanliness, Ltd. While patrons washed their bodies, the company's employees laundered and ironed the clients' clothes. The baths were heated and the floors warmed. They even had a jingle for their hypothetical company: "The baths of Caracalla where the rich and poor will walla."[55]

Although Dunn's theory about cultural factors was unique, his scientific approach supported a common trend in human studies at the time. He planned to acquire blood samples from a specific human group and test them for A, B, and O blood types and then compare the community's makeup to data procured from other populations, such as those with a similar descent or that resided nearby. Analysis of blood types and blood group genes gave scientists access to human genotypes, and was therefore "a powerful new tool for studying human biology and human evolution," according to Dunn.[56] Investigating gene frequencies among a population through the analysis of blood groups had started four decades prior to Dunn's research, and beginning in the 1920s, Germans had used the method to study races.[57] Furthermore, prior to Dunn's departure for Italy, one of his students at the Institute for the Study of Human Variation had produced promising results in support of the theory's validity. Mathematician and human geneticist L. D. Sanghvi, who specialized in the distribution of blood groups in India, had evidence that the caste system kept individuals of differing classes segregated to a greater degree than previously thought.[58] Religious rules among communities of the Braham caste in Bombay displayed different gene frequencies for blood groups. Sanghvi's results served as the topic of his dissertation, which he completed in 1954.[59] Sanghvi later had a successful career in human genetics. In 1973 he helped establish the Indian Society of Human Genetics and served as its first president. Today the society honors Sanghvi for his pivotal role in establishing human genetics as a discipline in India with a lifetime achievement award that carries his name.[60]

Blood group analysis developed into a popular research tool during the first half of the twentieth century, and enough data had been collected by 1954 for A. E. Mourant to publish his book *The Distribution of the Human Blood Groups*.[61] Mourant, who was one of a handful of prominent scientists performing studies along these lines at the time, went to the Institute for the Study of Human Variation in early 1953 as a part-time visiting professor of human genetics.[62] He resided in New York in the months before the Dunns went to Italy and helped Dunn learn about serology and blood typing.[63] Mourant, founder and director of the Blood Group Reference Laboratory of the Lister Institute, sent supplies for collecting and analyzing blood samples from his laboratory in London to Dunn in Italy.[64]

In preparation for the trip abroad, Dunn also utilized his connections within the genetics community to gather information about isolated populations in Italy. He contacted geneticist Salvador Luria at the University of Illinois. Luria had finished his medical degree at University of Turin in 1935 and eventually had to leave Italy because he was Jewish and the Italian government passed an anti-Semitic law. He relocated to France initially and then to the United States. He was a research assistant in the Department of Surgery at Columbia's College of Physicians and Surgeons for his first eighteen months in the country.[65] Luria told Dunn that World War II had increased intermarriage among Jews and adversely affected the isolation of these Jewish communities.[66] Dunn contacted several scholars at Italian and Hebrew universities who gave him information on the nuclear communities in Italy and aided the project not only while the Dunns resided in Rome but also after they returned to New York. For example, until 1931, Livio Livi at the University of Florence had performed investigations similar to those Dunn proposed in 1953. Dunn had read Livi's articles and sought his advice about Jewish communities in Italy, especially the one in Rome. Livi offered to help Dunn when he arrived and suggested that Dunn contact the president of the Jewish community in Rome. Dunn followed Livi's advice.[67]

Jews had lived in Rome since 167 BCE, making this community the oldest Jewish community with a continuous existence outside of Palestine. During the height of the Roman Empire (or, in other words, after the rise of Christianity in Rome), Roman laws segregated Jews from Christians for the first time. The initial laws restricted Jews professionally by dictating that they could not practice law or medicine or hold public office. Their situation deteriorated between the twelfth and nineteenth centuries. Jews were relegated to the ghetto in 1442, and walls were erected around the area as a result of a papal bull enacted by Pope Paul VI in 1555. The 1555 bull was extremely severe: a curfew confined Jews to the ghetto between ten at night and seven in the morning; they could not own property; the only professions accessible to Jews were collecting trash and trading secondhand goods; ceremonial and religious events were limited; and all Jews had to wear yellow hats. Even though subsequent popes lessened these rules over many decades, the career constraints kept the Roman Jews impoverished. Pope Urban VIII instated a rent freeze that had the long-term consequence of keeping Jews from moving outside the ghetto neighborhood even after the Papacy lost political power in 1870, thereby eradicating all laws restricting Jews. Some Jewish ghetto residents stayed in the neighborhood, choosing to live as their ancestors had and continuing the long legacy of segregation into the 1950s. Others had the means and motivation to move out of the ghetto.[68]

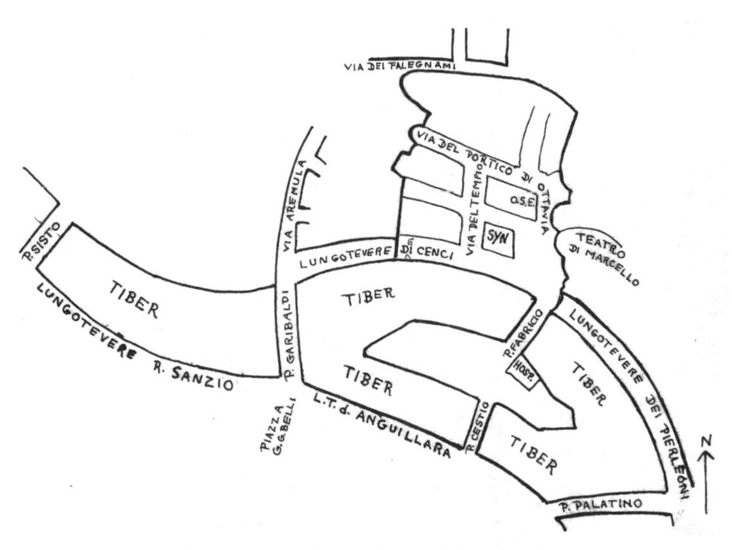

Figure 1. L. C. Dunn's hand-drawn map of the Jewish community's neighborhood in Rome. (Untitled map of Rome Jewish Community, Dunn Papers, series 5, box 31, Rome Jewish Community–Populations Statistics, Italy, American Philosophical Society, Philadelphia).

When the Dunn family was in Rome, the Jewish ghetto community resided in a small area of about ten square blocks east of the Tiber River, not far from Vatican City. This section of Rome has the lowest altitude, and before a dike was built for the Tiber River in the mid-nineteenth century, it was the unhealthiest section of the city.[69] Since the 1950s, few modern conveniences had been added to the four hundred- to five hundred-year-old buildings where most of the community's residents lived. Toilets were an open pipe emptying directly into the sewer, running water was fetched from a central location in the building or from the fountain in the square, and a kitchen was shared by several families.[70]

The facilities catering to the Jewish population were the heart of the community and were located on both sides of the Tiber River. On the southwest side of the river were the elementary school, orphanage, and children's clinic. To the north were the synagogue, welfare office, and adult clinic. The hospital was located between these two areas on an island. Portico d'Ottavia was the Jewish community's main street. Most communication among the community's members occurred at these locations.[71]

The Tiber River was also a geographical barrier separating what the Dunns referred to as the nuclear and peripheral communities. The nuclear community resided in the original ghetto north of the river and consisted of roughly four thousand people. An additional eight thousand people whose relatives had once been members of the Jewish ghetto community resided in the periphery around the ghetto. In both communities, Stephen interviewed people and Dunn collected biological samples. In total, 656 people participated: 267 from the ghetto, 172 from the other side of the Tiber River, and 217 within Rome and outside the other two areas.[72]

After arriving in Italy, it took the Dunns a few months to finalize some aspects for their field research, but they eventually worked out all the details. Stephen began his preliminary research on the community's history and social structure shortly after they settled into Rome, and according to Dunn he was "delighted & working hard" in interviewing residents and perusing archival documents.[73] Dunn collected blood in the clinic by approaching those there for vaccinations or to get their children examined so that they could attend school.[74] Each week Dunn divided his time between Rome, where he collected samples from the Jewish residents, and Milan, where he analyzed the samples for blood types. Stefano Fajrajzen, medical director of the Jewish Health Service, helped Dunn in Rome as a paid research associate of the Institute for the Study of Human Variation. Ruggero Ceppellini, chief of the hematology section of the Istituto Sieroterapico, worked with Dunn in Milan.

Ceppellini and Dunn collaborated for several years. Ceppellini was born in Milan and trained at the Lister Institute in London. Shortly after the Dunns left Italy, Ceppellini went at Columbia as a visiting investigator at the Institute for the Study of Human Variation. He remained in New York from 1954 to 1958. Dunn's perspective undoubtedly influenced the direction of Ceppellini's research, which increasingly included genetics and evolution.[75] Dunn and Ceppellini coauthored a couple of papers on their enclave investigations.[76] Ceppellini would eventually become a professor of medical genetics at the University of Turin in 1962.[77]

Dunn quickly learned that he had to obtain the trust of local community and organizational leaders as well as the cooperation of community members before he could begin the biological study. Diplomacy and material incentives ultimately gained the Dunns access to the people and information about them. Dunn wrote a letter to the ghetto community's leader in which he expressed his understanding of the need for confidentiality and promised to conduct his work with discretion.[78] In addition to obtaining the trust of authorities, he had to gain the confidence of the community members before they would submit to having blood drawn. Dunn quickly comprehended how best to get blood samples and retain his welcome in the community:

> *We're bringing in things needed in the community (milk powder, chocolate, vitamins, drugs) to distribute as incentives or more crudely in exchange for blood. If the rabbi ever finds out that we are getting "missionary rates" on some of these products, he'll throw me out, since there's nothing he fears so much as baptisms, mixed marriages & other "disruptive forces." We steer clear of religious leaders as much as possible & work with the doctors.*[79]

The Dunns also held monthly drawings in which the winning family received food packages valued at 6200 lire, which was equivalent to ten U.S. dollars.[80] To gain an understanding about the family size from which Dunn drew blood samples, the prizewinning families in January and February 1954 had eight and eleven members, respectively.[81]

Dunn, in one of his more candid appraisals of bartering material supplies for blood samples, said:

> *Great success! The first packet arrived this morning & this afternoon an obdurate family rolled up their sleeves. Its hard to say whether it was papa's cigarettes or la signora's & le signorine's "nailons" [nylons] that did the trick but the blood is in the (no, not bag) tubes.*[82]

Material goods successfully enticed the Jewish community's members to participate in the study because the Jews lived in squalor and usually begged for charity. In a 1958 article, Stephen noted that the poor Jews who lived within the ghetto were a distinct group separated from more wealthy Jews living outside, and that outsiders readily gave charity to ghetto residents. Stephen experienced this goodwill himself. More than once a passerby offered Stephen money, which he guessed resulted from his dependency on a wheelchair.[83] Stephen's comments on Jewish families' incomes provide insight into why exchanging food and other supplies for bodily serums would have been a reasonable approach in the early 1950s, even though it may seem exploitative by today's standards.

Indeed, ethical questions abound in reference to the Dunns' methodology of using material incentives to gain access to the poverty-stricken ghetto residents; however, their tactics should be viewed within a postwar context. The years following the war's end subjected many people in war-torn countries to hard times, in part because trade lines were not immediately reconnected. Dunn had lots of experience in sending publications and other supplies to scientists overseas in the years during and following World War II. Providing scientists in the Soviet Union with academic publications and research materials was one of the American-Soviet Science Society's greatest accomplishments. As president of the society, Dunn helped arrange these shipments through the State Department, the Smithsonian Institute, the Russian consulate in New York, and European embassies. He also personally sent reprints and scientific equipment in response to requests from geneticists in other war-torn countries, including Japan, England, and Norway.[84]

In fact, while in Rome, Dunn worked with Committee for American Remittances to Everywhere (CARE), a nonprofit organization that provided him with coffee, baby food, and high-protein food for the people in the Jewish community. A pamphlet summarized the agency's aim: "CARE is a key to open hearts locked by war, fear, mistrust and want . . ."[85] In a thank-you letter to one benefactor who provided CARE with items that were given to the Jewish community, Dunn noted that the donation was appreciated not only by the families receiving the goods but also by the scientists examining these families.

> *We are trying to find out what the factors are that have held such a community together for 2000 years and what have been the results of continued marriages between relatives, since nearly all marriages are contracted within the community. We take samples for analysis of blood, urine, and saliva, and as incentives to cooperate in our study we have given food, vitamines [sic] and medicines. This is where your packages did double duty both as relief and as a contribution to scientific research. They have been received with gratitude.*[86]

Dunn later wrote that diplomacy was the investigators' only means for getting results when working with human beings, noting that different situations require different methods of diplomacy.[87]

To truly appreciate the ethical circumstances surrounding the Dunns' biological and anthropological field research, it is important to elaborate upon the status of informed consent with respect to human experimentation in the early 1950s. Between the 1940s and 1970s, several medical incidences brought to light the violation of patients' rights, and, as a result, laws were passed to protect the patients. The Nuremberg Trial found guilty Nazi doctors who had performed a variety of experiments on human beings. In 1946 the United Nations General Assembly adopted the Nuremberg Code, which outlines four characteristics of consent—competent, voluntary, informed, and understanding—for anyone involved in medical research. The Nuremberg Code instituted an international law about informed consent and years later, in 1964, the medical community outlined its dictums in the Declaration of Helsinki. Both these have been influential to global acknowledgment of patients' rights.[88]

Contributing to the grand-scale adoption of informed consent rules were the handful of medical atrocities that came to light between the 1950s and 1970s. Two incidences of note are the Jewish Chronic Disease Hospital case and the Tuskegee Syphilis Study. Doctors at the Jewish Chronic Disease Hospital in Brooklyn, New York, introduced cancer into the bodies of twenty-two patients without informing them first. In 1966 the doctors were found guilty of fraud, deceit, and unprofessional conduct and given one year of probation. The Tuskegee Syphilis Study, which was conducted from the 1930s to 1970s, monitored but did not treat almost four hundred African American men who had contracted syphilis on their own accord. When this notorious case came to the public's attention in the early 1970s, it demonstrated that American medicine still failed to protect patients' rights. Both cases were under the jurisdiction of the United States Public Health Services (USPHS), and thus the incidences raised concerns among members of the PHS about moral responsibility in scientific investigations. One result was the National Research Act, which became law in 1974 and established the National Commission for the Protection of Human Subjects of Biomedical and Behavioral Research.[89]

Thus, while the Dunns were in Italy, the American medical community did not widely recognize the patient's rights to informed consent. Although the medical profession's situation should not be considered to have a direct correlation to the biological, anthropological, and behavioral sciences, it is a good gauge for the temporal development of ethical codes in the human sciences, including biology, anthropology, and psychology. Furthermore, it was Dunn's task to persuade members of the community to donate blood.[90] It is true that he used material incentives as tools of persuasion, but he also had gained the approval of the community's leaders and members. In other words, Dunn had the consent of his human research subjects, whether it was from the individual personally or from a child's parent.

The Dunn Family's Analysis of the Jewish Community of Rome

Both father and son gathered sufficient information to develop theoretical interpretations. Through his anthropological and sociological examination, Stephen found that the Jewish community of Rome had developed a close kinship and solid identity. Members of the Jewish community remained proud of and defined themselves by their ancestry and religion, yet the community's older inhabitants knew little about Judaism and religion actually meant very little to them. Stephen's historical data provided an explanation. Physical isolation from the mid-sixteenth to late nineteenth century impinged on communication of the Jews in Rome with other Europeans and Jews. Regulations restricted travel and censored printed materials, which resulted in a cultural stagnation. According to Stephen, after centuries of separation from others, the Jews had come to accept their mutual circumstances and find a comfortable way of life, but upon leaving the neighborhood, they were evaluated by different criteria and made to feel inferior. Most Jewish community members confined themselves to the local neighborhoods and Jewish agencies. Stephen drew a poignant conclusion about members of the Jewish community, which should be seen as rooted in his own experiences of being in and out of clinics for his physical condition. He reflected, "Once he goes outside that community, he is made to feel like a failure, a beggar, and of course no one likes this."[91]

The Roman Jews enjoyed a period of freedom that lasted from 1870 to 1938, ending when Benito Mussolini's Fascist government passed laws that once again segregated Jews from Christians. Children of the community started attending a separate school as a result of these anti-Semitic laws. When the Dunns lived in Rome, this school thrived with 500 students, 350 of whom were from the nuclear community, as were a number of the teachers. At the school, children from the Jewish community learned about their religion and culture and acquired a better understanding of the religious holidays their families continued to celebrate. Thus, the Dunn family witnessed a shift occurring among members of the Jewish community of Rome: the younger generation was learning about and embracing their religion, although most of their parents knew little about its rituals and their meanings.[92]

Louise's assessment of surnames highlighted another impact that wartime anti-Semitism had on the Jewish community of Rome. Twelve last names were common to 50 percent of the ghetto residents. The Dunns compared these twelve names to rosters of those deported during the Holocaust and found a similar frequency of the surnames among the more than two thousand people listed on the German rosters. Dunn noted that few returned after deportation—only 15 out of 2,091—and this greatly affected the community, leaving many widows and orphans. According to Dunn, the aftermath was still unfolding while he and his family lived in Rome.[93]

Stephen believed that one long-term consequence of the war was a decrease in the Jewish community's cohesion through an attempt at integrating them with the Roman population at large. Both the school and the orphanage that catered to the Jews gave children of the community a different outlook from that of their elders. Along with being more familiar with their religious roots than the older generation was, the students also received a higher level of education with the intention of putting them on par with other children. Thus, in its attempt to unite the Jewish community, the school was having a sociological effect that undermined that unity, according to Stephen. The orphanage, which was located across the Tiber River and quite a distance from the ghetto, provided a home to fifty children from the Jewish community. Since life at the orphanage was considerably better than in the ghetto, the worst punishment was threatening a child with being sent back to the ghetto. Those who grew up in the orphanage had a different demeanor from that of the ghetto's residents, causing most orphans to feel like outsiders if they returned to the neighborhood. After they left the orphanage, many orphans did choose to live in the ghetto area, but in comparison to those raised in the ghetto, the orphans did not feel the same intimate involvement with the community and its members.[94]

In terms of the enclave's marital habits, endogamy was evident within the nuclear community. The commonalities of surnames offered Stephen one piece of evidence that members of the Jewish ghetto tended to marry each other, and for additional proof he interviewed Jewish and non-Jewish Romans and perused community records. Dunn's biological results also provided evidence of intermarriages because they supplied solid proof of the community's uniqueness in comparison to surrounding populations. With the help of others, Dunn found two significant differences in gene frequency that separated the Jewish community from the larger population. There was a high frequency of blood type B within the Jewish community when compared with two non-Jewish population samples taken from Romans and Central Italians. Although European Jews tended to have a higher percentage of blood type B, the Jewish community of Rome had the highest frequency known to date. They also found more than 4 percent of r′ (Cde), an allele of the Rh system. The norm among Italians was under 1 percent. Dunn noted that these two unique gene frequencies, blood group B and the r′ (Cde) allele, could not be attributed to adaptive processes based solely on their evidence; however, he postulated that gene flow (i.e.,

Table 1. Dunn's results assessing gene frequencies blood type B and the Rh system allele r′ (Cde) of the Jewish Community in Rome and compared to two local populations

Group Analyzed	*Blood Type B*	*Rh system, r′ (Cde)*
Jewish Community	26.1%	4.5%
Romans	11.3%	-
Central Italians	10.6%	0.7%

migration), fluctuations in sampling, and natural selection could account for the results.[95]

The significantly higher frequencies of these two genes found among the Jewish community of Rome verified Dunn's hunch about the role that religion has on marital decisions, and thereby the impact that mating patterns have on the human gene pool. The almost threefold greater frequency for the blood group B gene led Dunn to write that "if these [two religious groups, Jews and Catholics] had been mixing with any regularity over a number of generations, these frequencies would have become similar."[96] And the almost fivefold difference in the Rh factor was further proof of the enclave's isolation. Taken together, these two pieces of data were "evidence that the efficacy of religious rules against intermarriage can be verified from biological evidence."[97]

Biological results were only part of the equation when studying human beings, according to Dunn. Due to human beings' long life cycles, it was not possible to detect how human populations had adapted biologically to the recent development of cities. Armed with social and cultural information, an investigator could augment biological data by giving it historical depth.

The data the Dunns gathered during their trip to Italy allowed them to construct a multidimensional view of the Jewish ghetto community in Rome; furthermore, it helped them contribute to the understanding of the development of human races. By forging a theoretical link between animal and human populations, Dunn argued that microevolution acted similarly on mice and human beings. Whereas biologists categorized animals into species and subspecies, human beings were divided into races and subraces. Moreover, whereas geographical environment dictated the microevolution of animals and humans, humans were social beings as well, and therefore both social and biological factors had to be considered.[98] Dunn also believed that mammalian geneticists' knowledge about developmental defects, whether triggered by environment or inheritance, could provide insights into inborn disorders in human beings. At this time, not all supporters of human genetics agreed that an understanding of human genetics could be attained from nonhuman subjects.[99]

The Dunns' data, however, did not provide conclusive results. Stephen defended his dissertation in 1956, but his dissertation committee told him that he needed more evidence. Dunn acknowledged that his son's dissertation was "thin" and thought the experience of revising it would be good for Stephen.[100] Stephen changed the focus of his dissertation, making it an exploration of the impact of ideology on culture. He added a chapter on Catholic ideology in Rome, investigated a second community, and successfully defended his dissertation three years later, in 1959. For his second group, Stephen analyzed Mexican Indians and the conquest of Mexico that occurred between 1516 and about 1650, and therefore he collected his data from historical documents. His fieldwork and conclusions on the Roman Jewish community filled almost twice as many pages as the Mexican Indian survey. Stephen hoped that he had demonstrated "the operation of a creative force in cultural evolution, not hitherto explicitly recognized—namely ideology, which, while still conceptually a part of culture, has achieved a considerable degree of autonomy from the cultural matrix out of which it sprang." He admitted that his dissertation lacked definitive answers but argued that he presented new methods and possible results for subsequent studies into the "The Influence of Ideology on Culture Change," as his dissertation was titled.[101]

Stephen's dissertation fits squarely into one avenue pursued by anthropologists during the 1950s: the extension of rural, small-town folk studies into peasant, small community studies. Two field studies undertaken in the 1920s serve as the origin of community studies. Robert Redfield, who trained in both sociology and anthropology at University of Chicago, performed a study in Mexico City. Lynd and his wife, Helen, who were personal friends of the Dunns, produced their groundbreaking study of Middletown that was based on residents in Muncie, Indiana. By the 1950s peasants and small communities had become a popular focus for anthropologists, and Conrad Arensberg, one contributor to the studies, oversaw the writing of Stephen's dissertation. Arensberg attempted the first community survey of a European population when he completed a study in Ireland during the 1930s. He made Europe a location worthy of research, but due to the war, such research was not executed until the 1950s, at which point several community studies were undertaken in Europe in addition to the Dunns' study in Italy.[102]

Dunn recognized that his scientific data also presented inconclusive results. He called himself an amateur when discussing his work in human and population genetics, and pointed out that his investigations had started some interesting new avenues for research but could not provide definitive answers. Dunn also investigated a second community: descendents of Africans from the Gullah tribe who lived on James Island off the shore of South Carolina. In the months after he returned from Italy, Dunn went to South Carolina to give four lectures titled "Heredity and the Community" when a few African American physicians presented him with the idea of conducting research on a Charleston community. Together they decided to proceed with the project, eventually incorporating researchers from

Howard University and the South Carolina Medical College. Those associated with the James Island study focused on variation by assessing gene frequencies with the ultimate goal of comparing this population to other African populations in the United States.[103] The James Island community had between 140 and 150 intermarrying families. Dunn and his coworkers collected complete hemoglobin data on forty-five of these families—a total of 280 people—and gathered incomplete samples on some of the remaining families.

Combining data from the Jewish and African communities, Dunn concluded that each group was biologically unique compared to the populations surrounding them. Dunn's biological data, however, only demonstrated that "Jews marry Jews, and Africans, Africans." More cross-disciplinary investigations had to be undertaken before further conclusions could be drawn.[104] Nonetheless, the family's preliminary research convinced Dunn that interdisciplinary studies drawing from anthropology, sociology, and genetics contributed valuably to the current understanding of human evolution. After enough data on gene frequencies could be collected about various populations inhabiting the earth, fruitful comparisons could be made and more profitable conclusions drawn. Researchers would later perform investigations on human populations as outlined by Dunn, but he discontinued his work along these lines.[105] It was 1961, one year before he retired and a few years after the Institute for the Study of Human Variation had ceased to exist. Although he continued to perform experiments after retiring, he concentrated on mice and not human beings.

After completing the fieldwork in Italy, Dunn used his newly gained perspective to warn people about the dangers of applied human genetics, previously known as eugenics. He was not opposed to weeding out deleterious genes from the human gene pool, but he did caution against implementing a program prematurely. For example, human genetics needed to be central to medicine before headway could be achieved. Dunn observed that genetics and medicine had been two separate entities in the past but that the gap between them was narrowing. Moreover, Dunn recognized that there existed a variety of ailments, each of which required a different (as opposed to a universal) solution. In terms of selection, abnormalities caused by a dominant gene could be eliminated in one generation with the correct diagnosis and program. Deformities due to recessive genes, however, were trickier to accurately diagnose and impossible to eliminate because many people lived normal and healthy lives while carrying the deleterious gene, which could be passed from parent to progeny. He suggested out-marriages as a method for decreasing the frequency of recessive ailments in isolated communities. Out-marriages would not remove defective genes but would spread the gene to a larger population, making it less likely for someone to inherit two recessive genes and therefore be born with a physical or mental disorder.[106]

As a culminating project, Dunn wrote *Heredity and Evolution in Human Populations*, a book readily accessible to nonscientific readers that was largely based on his knowledge of human genetics and serology and provides a summary of research accomplished through the Institute for the Study of Human Variation. Dunn presented a geneticist's view of human races and summarized the new approach that geneticists offered to the study of evolution as the difference between assessing genotypes and phenotypes. Geneticists and anthropologists examined genotypes, which deal with an individual's genetic constitution, when they performed surveys of blood groups because blood groups allowed investigators to learn about gene frequencies, which was the best approach for understanding human evolution. Older methods of anthropometry depended on descriptions and measurements of the human body and thus analyzed phenotypes. Phenotypes, the physical features based on a person's genetic makeup, are affected by environmental factors to a greater degree than genotypes.[107] It was the geneticist's task to examine the relationship between genotypes and phenotypes, according to Dunn.[108]

After explaining to his reader the benefits of blood group antigen research for learning about human evolution and variation, Dunn discussed the research that had been performed on isolated human populations, mentioning six communities in total. Five of these, including the Jewish community of Rome and the Gullah tribe of James Island, were analyzed by members of the Institute for the Study of Human Variation.[109] Ultimately, these studies reinforced Dunn's view that social and cultural factors implemented by choice or by force have kept human communities segregated for centuries. These cultural barriers succeeded at creating biological variation among the world's population. Even as civilization and urbanization decreased the number of geographical barriers, social isolation continued to impede integration. Therefore, evolutionary forces that did not act on other living organisms affected human beings.[110]

Dunn commented on the book's value, as did its reviewers. According to Dunn, readers "who were looking not for easy answers or for final ones but for promising ones, that is, those that presented new views or opened new ways of study" were drawn to the book. Moreover, he had learned a valuable lesson over his lifetime. Genetics was not the exact science its proponents claimed it to be when he started his studies in the

1910s, but rather an explorative, maturing science, and the discipline's future looked optimistic.[111] Reviewers liked the book and found few flaws with it. Human geneticist Bentley Glass commented on Dunn's prose of "perfect lucidity in handling the subject" and his scientific competence to produce "a comprehensive, balanced view." Moreover, Glass hailed Dunn for his ethical stance because Dunn showed "evidence of a social conscience well-disciplined by respect for scientific evidence." It should be noted that Dunn drew on Glass's research and even extensively quoted Glass in the book.[112] Glass therefore must be seen as an informed and biased evaluator. Nevertheless, more than fifteen years after Dunn wrote *Heredity and Evolution in Human Populations*, it was "still looked upon as an indispensable source of scientific data on race."[113]

Overall, Dunn published more studies on the Roman Jews than on the James Island Africans, which can be explained by one or both of the following reasons: First, the Institute for the Study of Human Variation was nearing its end around the time that Dunn was processing the James Island information, and shortly after the institute closed, he submitted the blood data collected under its auspices on the Roman Jews and African Americans to the University Microfilm Bureau in Ann Arbor, Michigan.[114] This indicates that he chose to return to his specialty: developmental genetics in mice. Second, Dunn continued to publish articles and give lectures on the Roman Jews into the 1960s in connection with his son, which reflects Dunn's concern for Stephen's career. Although Stephen held a couple of short academic appointments, he did not enjoy the guarantee of permanent, ongoing employment.

Dunn and Stephen remained close until Dunn's death. After retiring in 1962, Dunn often traveled to visit Stephen and immensely enjoyed their time together, as his following remark attests: "I believe I have more pleasure in visiting him and in the more or less continuous discussions which occupies us when together than in any other activity of any kind." Dunn was also proud that Stephen had followed in his footsteps. "Stephen finished his lectures at Monterey [Institute of Foreign Studies] just before we arrived. I have read them—fourteen of them each beginning with a title & an outline and ending with a bibliography—the son fulfills the ambition of the father."[115] In 1974, five days before Stephen's forty-sixth birthday, Dunn passed away unexpectedly at the age of eighty. He was survived by his wife and two sons and deeply missed.[116] Stephen Dunn passed away in 1999.

Conclusion

Dunn, who conceived and initiated the family's investigative vacation, united several interests through his study. Science undoubtedly fueled the research, but only to a certain extent because social and political issues underlined his scientific investigation. The nature of the Dunn family's research trip demonstrates the overlap between Dunn's personal and professional lives and shows how he was able to bring together his motivations, experiences, and opinions that had been developing over a thirty-year span. Ultimately, Dunn wished to learn more about the diversity of human races and to clarify the definition of race. Although Dunn did not consider Jews to be a race, he believed that many people in the world misunderstood the differences between race, religion, cultural group, and language. He initially responded to discriminatory laws by attacking racial prejudices, and later started an interdisciplinary institute to study human populations. He went into the field to collect data on human beings in an attempt to study human evolution and produced fruitful yet preliminary conclusions about the evolutionary history of human beings.

Informed by his biological data, Louise's demographic survey, and Stephen's social, economic, and historical information, Dunn wrote in 1959 that "isolating factors in human populations may be geographical or social." He also defined the aspects that create barriers: "Differences in language, religion, or custom may act as social impediments to mating." Then he added:

> *In some societies race prejudice—a social acquirement—may be an isolating factor, as in the southern United States. The Roman Jewish community has been a social isolate formed under the influence of a variety of historical factors both religious and economical.*[117]

As noted in his statement, Dunn linked the situations facing the Jewish community living in Rome and the African American community on James Island. Furthermore, he argued that religious and racial ancestry influenced evolution because cultural norms and social prejudices guided people's marriage decisions and thereby affected their progeny's biological constitution.[118]

Even though Dunn stated that he did not set out to investigate a Jewish group in Italy, he eventually used his research on two communities that suffered lengthy periods of forced isolation to make a statement about discrimination. His study of the African American community living in the southern United States addressed racial prejudices, segregation, and the inferior treatment of African Americans. His study of the Jewish community of Rome focused on a religious group that had been persecuted in a Fascist country during World War II. The Dunn family gathered biological and cultural data that would allow Dunn to counter biases affecting African Americans and those held over from Nazism. In his 1961 speech as president of the American Society of Human Genetics, Dunn used the pulpit to examine historical reasons for why human genetics developed slowly internationally, concluding that various forms of scientific discrimination were the main culprits in the United States, England, Germany, and Russia. His advice: human geneticists must address both the biological and social problems because only those knowledgeable of the science can properly critique and give advice on the social applications of their discipline.[119]

Notes

1. L. C. Dunn, "What Did We Want to Find Out?" (symposium lecture, 1968), 3, Dunn Papers, series 5, box 32, Rome Jewish Community–"Man in the City," American Philosophical Society, Philadelphia. "The Roman Campaign of '53–'55" is the Dunn family's phrase used to refer to their trip to Italy and the work resulting from their time there.
2. L. C. Dunn, "The Reminiscences of Leslie Clarence Dunn," *Oral History Collection of Columbia University*, interviews by Saul Benison, 1958–1960, microfilm, transcript, 900.
3. Each member of the Dunn family, except for L. C. Dunn, will be referred to by first name in this article.
4. L. C. Dunn, "Reminiscences of Dunn," 904 (see n. 2).
5. Theodosius Dobzhansky to L. C. Dunn, 22 October 1953, Dunn Papers, series 1, box 6, Theodosius Dobzhansky-Dunn Correspondence #7, 1950–1953, American Philosophical Society, Philadelphia; and L. C. Dunn, "A Genetical Study of a Jewish Community, the Old Ghetto in Rome," n.d., 9, Dunn Papers, series 5, box 32, Rome Jewish Community–"A Genetical Study of a Jewish Community, the Old Ghetto in Rome," n.d., American Philosophical Society, Philadelphia. For a good overview of their study and its results, see L. C. Dunn and Stephen P. Dunn, "The Jewish Community of Rome," *Scientific American* 196 (March 1957): 118–24. The term "ghetto" was originally used to designate these locations in Italy where Jews lived. It has since developed a more widespread definition to designate a dilapidated area inhabited by a minority or several minorities. *Oxford English Dictionary Online*, s.v. "ghetto," http://dictionary.oed.com (accessed May 26, 2009).
6. L. C. Dunn, "Heredity and the Study of Small Communities," (Madison Lecture Series, 1 April 1955), 7, Dunn Papers, series 1, box 7, L. C. Dunn–"Heredity and the Study of Small Communities," 1955, American Philosophical Society, Philadelphia. Many other people helped the Dunn family. See the following for the most comprehensive list of collaborators: L. C. Dunn, "Genetical Study of a Jewish Community," n.d., Acknowledgments [22–23] (see n. 5).
7. L. C. Dunn and Stephen Dunn, "Jewish Community of Rome," 128 (see n. 5).
8. Dunn Papers, series 1, box 10, Emergency Committee in Aid of Displaced German Scholars, 1933, American Philosophical Society, Philadelphia; Dunn

Papers, series 1, box 11, Emergency Committee in Aid of Displaced German Scholars–Minutes, 1933, American Philosophical Society, Philadelphia.

9. Dunn Papers, series 1, boxes 10–13, Emergency Committee in Aid of Displaced German Scholars, American Philosophical Society, Philadelphia; Emergency Committee in Aid of Displaced Foreign Scholars, New York Public Library–Manuscripts and Archives, box 167. Executive Committee members bequeathed the Emergency Committee's records to the New York Public Library upon the organization's cessation. Dunn Papers, series 1, box 5, Columbia University–Faculty Fellowship Fund, American Philosophical Society, Philadelphia.
10. L. C. Dunn, "Reminiscences of Dunn," 877 (see n. 2).
11. L. C. Dunn and Theodosius Dobzhansky, *Heredity, Race and Society*, 2nd ed. (1946; 2nd ed., New York: New American Library of World Literature, 1952): 40–42, 114–17.
12. L. C. Dunn and Dobzhansky, *Heredity, Race and Society* (see n. 11); L. C. Dunn, *Race and Biology* (Paris: UNESCO, 1951); UNESCO, *What Is Race?: Evidence from Scientists* (Paris: UNESCO, 1952); and UNESCO, *Race and Science: The Race Question in Modern Science* (New York: Columbia University Press, 1961).
13. L. C. Dunn and Dobzhansky, *Heredity, Race and Society*, 118 (see n. 11). The authors gave a similar definition later in the book: "Races are populations which differ in the relative commonness of some of their genes" (L. C. Dunn and Dobzhansky, *Heredity, Race and Society*, 124).
14. L. C. Dunn, *Race and Biology*, 15 (see n. 12).
15. Ibid., 14.
16. L. C. Dunn and Dobzhanksy, *Heredity, Race and Society*, 130 (see n. 11).
17. L. C. Dunn, *Race and Biology*, 25 (see n. 12).
18. L. C. Dunn to Walter Landauer, 27 June 1943, 6 March 1944, 25 July 1944, 31 August 1944, and 8 September 1944, Dunn Papers, series 1, box 19, Walter Landauer, 1943–1944, American Philosophical Society, Philadelphia. Bob was home from 31 August to 12 September 1944.
19. Robert S. Lynd to Frank D. Fackenthal, 21 September [1940], and Fackenthal to Lynd, 23 September 1940, central files, box 378, folder 1, Robert S. Lynd, June 1934 to July 1935 and March 1937 to December 1940, University Archives and Columbiana Library, Columbia University, New York.
20. Frank D. Fackenthal to L. C. Dunn, 4 March 1940 and 4 October 1943, Dunn Papers, series 1, box 13, Frank D. Fackenthal (Columbia University), 1934–1949, American Philosophical Society, Philadelphia. L. C. Dunn was designated executive officer of the Zoology Department at a faculty meeting on 4 March 1940 for the period 1 July 1940 to 30 June 1943. Fackenthal informed Dunn on 4 October 1943 that he was appointed "Executive Officer of the Department of Zoology for three years from July 1, 1943."
21. Theodosius Dobzhansky to L. C. Dunn, 2 October 1946 and 23 January 1947, Dunn Papers, series 5, box 6, Theodosius Dobzhansky–Dunn Correspondence #5, 1946–1947, American Philosophical Society, Philadelphia.
22. L. C. Dunn to Warren Weaver, 12 February 1947, Rockefeller Foundation, RF 1.2 200D, box 174, folder 1599, American-Soviet Science Society, 1946–1958, Rockefeller Archive Center, Sleepy Hollow, NY. According to a letter from L. C. Dunn to Walter Landauer, Dunn and his family were still in the United States as late as March 1947 (L. C. Dunn to Walter Landauer, 30 March 1947, Dunn

Papers, series 1, box 19, Walter Landauer, 1947, American Philosophical Society, Philadelphia).

23. L. C. Dunn, "Reminiscences of Dunn," 895–96 (see n. 2); and Gunnar Dahlberg, *Race, Reason, and Rubbish: An Examination of the Biological Credentials of the Nazi Creed*, trans. Lancelot Hogben (1942; repr., London: George Allen & Unwin, 1943).
24. L. C. Dunn, "Reminiscences of Dunn," 893–95 (see n. 2).
25. Ethel Dunn, "Stephen P. Dunn," *Anthropology Newsletter* 40, 6 (June 1999): 48.
26. L. C. Dunn to Landauer, 1 February 1947 (see n. 22).
27. L. C. Dunn to Walter Landauer, 21 March 1929, Dunn Papers, series 1, box 18, Walter Landauer, 1929, American Philosophical Society, Philadelphia.
28. Louise Porter Dunn to L. C. Dunn, n.d., Dunn Papers, series 1, box 9, Louise Porter Dunn, 1927, n.d., American Philosophical Society, Philadelphia; and L. C. Dunn to Walter Landauer, 6 June 1934, Dunn Papers, series 1, box 19, Walter Landauer, 1934, American Philosophical Society, Philadelphia.
29. L. C. Dunn to Walter Landauer, 18 September and 18 October 1930, Dunn Papers, series 1, box 18, Walter Landuaer, 1930, American Philosophical Society, Philadelphia; Dunn to Landauer, 6 May and 25 June 1932, Dunn Papers, series 1, box 18, Walter Landuaer, 1932, American Philosophical Society, Philadelphia.
30. L. C. Dunn to Walter Landauer, 27 June, 13 August, and 27 September 1943, Dunn Papers, series 1, box 19, Walter Landauer, 1943, American Philosophical Society, Philadelphia. Dunn told Landauer about Stephen's hand operation on 27 June 1943 and his troubles after the surgery on 13 August 1943, and wrote the remark quoted here on 27 September 1943.
31. L. C. Dunn to Walter Landauer, 12 June 1941, and L. C. Dunn to Landauer, 23 September 1945, Dunn Papers, series 1, box 19, Walter Landauer, 1940s, American Philosophical Society, Philadelphia. Stephen was accepted at Lincoln School of Columbia University in 1941 (Dunn to Landauer, 12 June 1941). Stephen started at Columbia in 1945 (Dunn to Landauer, 23 September 1945).
32. L. C. Dunn to Walter Landauer, 21 May 1970, Dunn Papers, series 1, box 20, Walter Landauer, 1970, American Philosophical Society, Philadelphia.
33. L. C. Dunn to Walter Landauer, 6 October 1943 and 25 July 1944, Dunn Papers, series 1, box 19, Walter Landauer, 1943 and 1944, American Philosophical Society, Philadelphia. Louise needed L. C. Dunn to help with Stephen, so he could not leave New York as planned.
34. Garland E. Allen, "The Eugenics Record Office at Cold Spring Harbor, 1910–1940: An Essay in Institutional History," *Osiris* 2 (1986): 251.
35. For information about the languages that L. C. and Stephen Dunn knew, see L. C. Dunn to Landauer, 15 September 1934 (see n. 28). About Steven learning Russian, see Theodosius Dobzhansky to L. C. Dunn, 6 August 1943, Dunn Papers, series 1, box 6, Theodosius Dobzhansky–Dunn Correspondence #4, 1943–1945, American Philosophical Society, Philadelphia.
36. Dobzhansky to L. C. Dunn (see n. 35).
37. Thomas Hylland Eriksen and Finn Sivert Nielsen, *A History of Anthropology* (London: Pluto Press, 2001), 159.
38. L. C. Dunn to James Lamar Weygand, n.d., Dunn Papers, series 1, box 4, Coalbin Press, 1956–1958, n.d., American Philosophical Society, Philadelphia; "Autobiographical Resume," 2, Dunn Papers, series 1, box 7, L. C. Dunn–

Autobiographical data, n.d., American Philosophical Society, Philadelphia. The Stephen P. Dunn of this article is not the Pulitzer Prize–winning poet Stephen Dunn. The other books published by the Dunns through Coalbin Press are as follows: L. C. Dunn, *The Coalbin Press Anthology* (1933; 25 copies); Robert, Stephen, Louise, and Leslie Dunn, *Other Countries* (1942; 40 copies); L. C. and S. P. Dunn, *Prose and Verse* (1950; 40 copies); and *Some Watercolors from Venice* (1956; 500 copies).

39. L. C. Dunn to Walter Landauer, 9 April 1955, Dunn Papers, series 1, box 20, Walter Landauer, 1955, American Philosophical Society, Philadelphia.
40. L. C. Dunn to Walter Landauer, 26 September 1956, Dunn Papers, series 1, box 20, Walter Landauer, 1956, American Philosophical Society, Philadelphia.
41. Ethel Dunn, "Stephen P. Dunn," 48 (see n. 25).
42. L. C. Dunn to Walter Landauer, 21 October 1942, Dunn Papers, series 1, box 19, Walter Landauer, 1942, American Philosophical Society, Philadelphia.
43. L. C. Dunn to Walter Landauer, 18 January 1952, Dunn Papers, series 1, box 20, Walter Landauer, 1952, American Philosophical Society, Philadelphia.
44. Stephen P. Dunn, "The Influence of Ideology on Culture Change: Two Test Cases" (diss., Columbia University, 1959).
45. L. C. Dunn to Walter Landauer, 1 February 1947 and 13 October 1948, Dunn Papers, series 1, box 19, Walter Landauer, 1947 and 1948, American Philosophical Society, Philadelphia.
46. L. C. Dunn, *Race and Biology*, 22–26 (see n. 12); and L. C. Dunn and Dobzhansky, *Heredity, Race and Society*, 130 (see n. 11).
47. L. C. Dunn and Dobzhansky, *Heredity, Race and Society*, 105–115 (see n. 11); ibid., 121–35; and L. C. Dunn, *Race and Biology*, 36–37 (see n. 12).
48. L. C. Dunn, "Heredity and the Study of Small Communities," 5–6 (see n. 6).
49. L. C. Dunn, "Reminiscences of Dunn," 904 (see n. 2).
50. Ibid., 907–8. The Anav family (which is sometimes spelled "Anau") had been living in Rome for many generations. Jewish Virtual Library, a division of the American-Israeli Cooperative Enterprise, http://www.jewishvirtuallibrary.org/jsource/biography/Anav.html (accessed November 21, 2008).
51. L. C. Dunn, "Reminiscences of Dunn," 911 (see n. 2).
52. L. C. Dunn to Walter Landauer, 11 April 1953, Dunn Papers, series 1, box 20, Walter Landauer, 1953, American Philosophical Society, Philadelphia.
53. R. B. Shipley to L. C. Dunn, 9 April 1953, and L. C. Dunn to Secretary of State John Foster Dulles, n.d. (written in response to Shipley's letter dated 9 April 1953), Dunn Papers, series 1, box 8, L. C. Dunn–Passport Correspondence, 1953, American Philosophical Society, Philadelphia.
54. L. C. Dunn to Landauer, 16 May 1953 (see n. 52). L. C. Dunn received his passport the day before writing this letter to Landauer.
55. Untitled letter from the Dunn family, 15 November [1953], Dunn Papers, series 5, box 31, Rome Jewish Community–Correspondence, 1955–1956, n.d., American Philosophical Society, Philadelphia. Handwritten at the top of the letter is "11/15." This would have been 1953.
56. L. C. Dunn, "Heredity and Environment" (lecture, Frontiers of Knowledge, December 16, 1952–January 7, 1953, New York University), 17, Dunn Papers, series 1, box 7, L. C. Dunn– "Heredity and Environment," 1953, American Philosophical Society, Philadelphia. L. C. Dunn gave this lecture a few months

before James Watson and Francis Crick published their celebrated article in April 1953 describing the structure of deoxyribonucleic acid (DNA). Blood and DNA have since become the two main biological components for research into genetics and evolution. James D. Watson and Francis H. C. Crick, "Molecular Structure of Nucleic Acids: A Structure of Deoxyribose Nucleic Acid," *Nature* 171 (April 25, 1953): 737–38.

57. Daniel Kevles, *In the Name of Eugenics: Genetics and the Uses of Human Heredity* (Cambridge, MA: Harvard University Press, 1985): 195.
58. Notes in Warren Weaver's diary, 7 February 1951, Rockefeller Foundation, RF 1.1 200D, box 132, folder 1633, Columbia University–Genetics, 1950–1952, Rockefeller Archive Center, Sleepy Hollow, NY.
59. L. C. Dunn, "Heredity and Environment," 17–18 (see n. 56); L. C. Dunn to John A. Krout, 4 June 1956, central files, box 382, folder 5, L. C. Dunn, 1954–1957, University Archives and Columbiana Library, Columbia University, New York; L. D. Sanghvi, "Comparison of Genetical and Morphological Methods for a Study of Biological Differences" (diss., Columbia University, 1954); and Institute for the Study of Human Variation–Report on Research Activities, 30 November 1956, Rockefeller Foundation, RF 1.1 200D, box 132, folder 1635, Columbia University–Genetics, 1955–1958, Rockefeller Archive Center, Sleepy Hollow, NY.
60. L. D. Sanghvi, ed., *Human Population Genetics in India: Proceedings of the First Conference of the Indian Society of Human Genetics* (New Delhi: Sangam Press, 1974), 289. The history of the Indian Society of Human Genetics (ISHG) and a description of the Dr. L. D. Sanghvi Oration Award are available on the ISHG Web site, http://www.ishg.in/about_ishg.htm and http://www.ishg.in/drldsanshavi_award.htm (accessed May 26, 2009).
61. Spencer Wells, *The Journey of Man: A Genetic Odyssey* (Princeton, NJ: Princeton University Press, 2002), 16. Wells stated that Mourant's book (A. E. Mourant, *The Distribution of the Human Blood Groups* [Oxford: Blackwell, 1954]) inaugurated the modern era of human genetics.
62. Resolution RF 57036, 21 February 1957, Rockefeller Foundation, RF 1.1 200D, box 132, folder 1631, Columbia University–Genetics, 1942–1947, Rockefeller Archive Center, Sleepy Hollow, NY.
63. L. C. Dunn to Landauer, 19 January and 25 February 1953 (see n. 52).
64. L. C. Dunn to Arthur Mourant and Philip Levine, telegrams, 25 January 1954, Dunn Papers, series 5, box 31, Rome Jewish Community–Correspondence, 1954, American Philosophical Society, Philadelphia.
65. "Detail of Information," 15 November 1940, Rockefeller Foundation, RF 1.1 200D, box 133, folder 1641, Columbia University–Salvador Luria (Refugee Scholar, Surgery), 1940–41, Rockefeller Archive Center, Sleepy Hollow, NY.
66. Salvador Luria to L. C. Dunn, 13 March 1953, Dunn Papers, series 5, box 31, Rome Jewish Community–Correspondence, 1953, American Philosophical Society, Philadelphia.
67. Elizabeth Goldschmidt to L. C. Dunn, 28 January 1954 (see n. 64); and L. C. Dunn to Livio Livi, 13 January and 9 February, and 5 February 1954, and Livi to L. C. Dunn, 31 January 1953, Dunn Papers, series 5, box 31, Rome Jewish Community–Populations Statistics, Italy, American Philosophical Society, Philadelphia.

68. Stephen Dunn, "Influence of Ideology," 217–37 (see n. 44) [Chapter 8 of Stephen's dissertation is titled "The Jewish Community of Rome: Historical Survey."]; and Stephen P. Dunn, "An Outsider Visits the Roman Ghetto: The Past Lives On," *Commentary* 25 (1958): 130–40, 131–33.
69. Stephen Dunn, "Outsider Visits the Roman Ghetto," 130 (see n. 68).
70. Stephen Dunn, "Influence of Ideology," 115–17 (see n. 44).
71. Stephen Dunn, "Influence of Ideology," 122–23, 139–53 (see n. 44); and L. C. Dunn and S. P. Dunn, "The Ancient Jewish Community of Rome," n.d., 5–7, Dunn Papers, series 5, box 32, Rome Jewish Community–"The Ancient Jewish Community of Rome," American Philosophical Society, Philadelphia. The most recent date on this manuscript is 1967, indicating that this manuscript was written in or after 1967.
72. Stephen Dunn, "Influence of Ideology," 114–15 (see n. 44); and L. C. Dunn and S. P. Dunn, "Ancient Jewish Community of Rome," 5–6 (see n. 71).
73. L. C. Dunn to Walter Landauer, 24 November 1953 (see n. 52); and L. C. Dunn to Steffano Fajrajzen of the Jewish Health Service, 4 November 1953 (see n. 66).
74. L. C. Dunn, "Reminiscences of Dunn," 912–13 (see n. 2).
75. L. C. Dunn, "Reminiscences of Dunn," 919 (see n. 2); Fritz H. Bach, "Ruggero Ceppellini (1917–1988)," *Immunology Today* 9 (1988): 335–37; W. F. Bodmer, "In Memoriam: Ruggero Ceppellini (1917–1988)," *Immunogenetics*, 29 (1989): 145–47; and Jon J. Van Rood and D. Bernard Amos, "In Memoriam: Ruggero Ceppellini, 1917–1988," *Human Immunology* 23 (September 1988): 1–3.
76. L. C. Dunn and Stephen Dunn, "Jewish Community of Rome," 124 (see n. 5); L. C. Dunn, "Genetical Study of a Jewish Community," n.d., Acknowledgments [22–23] (see n. 5); L. C. Dunn, "Biological Structure of a Small Community as Revealed by Blood Groups Genes," n.d., series 5, box 31, Rome Jewish Community–Notes, American Philosophical Society, Philadelphia [Based on the wording, this proposal was written before the work was done.]; L. C. Dunn to secretary of Institute for the Study of Human Variation, 22 November 1953 (see n. 66); L. C. Dunn, Ruggero Ceppellini, and Mario Turri, "An Interaction between Alleles and the Rh Locus in Man Which Weakens the Reactivity of the Rh_0 Factor (D^u)," *Proceedings of the National Academy of Sciences USA* 41 (1955): 283–88; and William S. Pollitzer, R. M. Menegaz-Bock, Ruggero Ceppellini, and L. C. Dunn, "Blood Factors and Morphology of the Negroes of James Island, Charleston, S.C.," *American Journal of Physical Anthropology* 22 (1967): 393–98.
77. Bach, "Ruggero Ceppellini (1917–1988)," 335 (see n. 75).
78. L. C. Dunn to Mr. Presidente, n.d., Dunn Papers, series 5, box 31, Rome Jewish Community–Notes, American Philosophical Society, Philadelphia.
79. L. C. Dunn to Walter Landauer, 24 November 1953 (see n. 52).
80. The conversion rate, according to Stephen P. Dunn, was 620 lire to 1 U.S. dollar.
81. L. C. Dunn, "A Biological Study of the Comunita Ebraica di Roma," n.d. (see n. 78).
82. L. C. Dunn to Marvin L. Edwards, 13 March 1954, Dunn Papers, series 1, box 9, Marvin L. Edwards #4, 1953–1954, American Philosophical Society, Philadelphia.
83. Stephen Dunn, "Outsider Visits the Roman Ghetto," 134–35 (see n. 68).
84. "Minutes of the Meeting of the Science Committee of the National Council of American-Soviet Friendship," 12 March [1945], 1, Dunn Papers, series 1, box

2, American-Soviet Science Society, 1944–46, n.d., American Philosophical Society, Philadelphia; L. C. Dunn to S. Gershenson, 27 July 1945, and L. C. Dunn to Annette Terzian, L. C. Dunn's secretary to H. W. Dorsey of the Smithsonian Exchange Service, 12 November 1945, Dunn Papers, series 1, box 25, USSR Correspondence with Geneticists, 1944–45, American Philosophical Society, Philadelphia; and Y. Ogura to L. C. Dunn, 3 April 1948, and L. C. Dunn's reply, 27 May 1948, Dunn Papers, series 1, box 23, Y. Ogura, 1948, American Philosophical Society, Philadelphia.

85. Assorted letters, November 1953–4 February 1954, Dunn Papers, series 1, box 3, C.A.R.E. (Committee for American Remittances to Everywhere), 1953–54, American Philosophical Society, Philadelphia.
86. L. C. Dunn to Mrs. Olen, n.d. (see n. 85).
87. L. C. Dunn, "Big and Little Populations: An Amateur's Excursion," *American Naturalist* (May–June 1961): 135.
88. George J. Annas, Leonard H. Glantz, and Barbara F. Katz, *Informed Consent to Human Experimentation: The Subject's Dilemma* (Cambridge, MA: Ballinger Publishing Company, 1977): 6–9; and Ruth R. Faden and Tom L. Beauchamp, *A History and Theory of Informed Consent* (New York: Oxford University Press, 1986): 151–57.
89. Faden and Beauchamp, *History and Theory of Informed Consent*, 161–67 (see n. 88).
90. Dunn, "Reminiscences of Dunn," 911 (see n. 2).
91. Stephen Dunn, "Outsider Visits the Roman Ghetto," 140 (see n. 68).
92. Stephen Dunn, "Influence of Ideology," 234–36 (see n. 44); and Stephen Dunn, "Outsider Visits the Roman Ghetto," 137–38 (see n. 68).
93. L. C. Dunn, "Genetical Study of a Jewish Community," n.d., 13–15 (see n. 5); Stephen Dunn, "Influence of Ideology," 237 (see n. 44); and L. C. Dunn to Mrs. Olen, n.d. (see n. 86). L. C. Dunn's figures may not be entirely accurate. According to Susan Zuccotti, the figure of 2,091 is the highest estimate. Other estimates of the number of Jews deported from the Roman ghetto community were 1,727 and 1,739, and these smaller estimates state that the total who returned was 105 or 114. Zucchoti notes that the statistical differences result from women being listed twice under both their married and maiden names and that some may have been arrested outside of Rome. Susan Zuccotti, *The Italians and the Holocaust: Persecution, Rescue, and Survival* (New York: Basic Books, 1987): 195, footnote 21.
94. Stephen Dunn, "Outsider Visits the Roman Ghetto," 137–38 (see n. 68); and Stephen P. Dunn, "Influence of Ideology," 145–52 (see n. 44).
95. L. C. Dunn, "Genetical Study of a Jewish Community," n.d., 16–20 (see n. 5); and L. C. Dunn and S. P. Dunn, "Ancient Jewish Community of Rome," 7–9 (see n. 71).
96. L. C. Dunn, "Reminiscences of Dunn," 914 (see n. 2).
97. Ibid., 914–15.
98. L. C. Dunn and S. P. Dunn, "Ancient Jewish Community of Rome," n.d., 3–4 (see n. 71); L. C. Dunn, "What Did We Want to Find Out?" (see n. 1); L. C. Dunn, "Heredity and the Study of Small Communities," 1–7 (see n. 6).
99. L. C. Dunn, "Problems in Genetics in Mammals," *Scientia Rivista di Scienza* 91 (1956): 51–56. According to Theodosius Dobzhansky, Warren Weaver of the

Rockefeller Foundation discounted *Drosophila* and mice as important to human genetics, as did James V. Neel, who believed that scientists could only learn about human genetics by working with human beings. Theodosius Dobzhansky to L. C. Dunn, 14 August 1955, Dunn Papers, series 1, box 6, Theodosius Dobzhansky–Dunn Correspondence #8, 1954–55, American Philosophical Society, Philadelphia.

100. L. C. Dunn to Walter Landauer, 7 April 1956 (see n. 40).
101. Stephen Dunn, "Influence of Ideology," 265–66 (see n. 44).
102. Sydel Silverman, "Bringing Anthropology into the Modern World," in *One Discipline, Four Ways: British, German, French, and American Anthropology*, ed. Fredrik Barth, Andre Gingrich, Robert Parkin, and Sydel Silverman, 292–309 (Chicago: University of Chicago Press, 2005).
103. L. C. Dunn's handwritten notes, n.d., Dunn Papers, series 5, box 31, James Island, South Carolina–Genetics Material, American Philosophical Society, Philadelphia.
104. L. C. Dunn, "Big and Little Populations," 136 (see n. 87); and L. C. Dunn, "Heredity and the Study of Small Communities," 2 (see n. 6).
105. One notable example was geneticist James V. Neel, who conducted his first field research on an Amerindian tribe with an anthropologist in 1962. Later, in 1966, he teamed up with anthropologist Napoleon A. Chagnon to study the Yanomamö, which has since become a well-known anthropological case study. Dunn and Stephen's results pale in comparison to those achieved by Neel and Chagnon. James V. Neel, *Physician to the Gene Pool: Genetic Lessons and Other Stories* (New York: John Wiley & Sons, 1994): 117–34; and Napolean A. Chagnon, *Yąnomamö*, 4th ed. (1968; 4th ed., Fort Worth, TX: Harcourt Brace Jovanovich College Publishers, 1992).
106. L. C. Dunn, "Lecture III: The Social Direction of Human Evolution," in "A Synopsis of Three Lectures on Heredity and the Human Community," *Bulletin of the Wagner Free Institute of Science* 30 (May 1955): 23–28; L. C. Dunn, "The Prospects for Genetic Improvement," *Eugenics Quarterly* 3 (1956): 37; and L. C. Dunn, "Cross Currents in the History of Human Genetics," *American Journal of Human Genetics* 14 (1962): 10.
107. L. C. Dunn, *Heredity and Evolution in Human Populations*, 2nd ed. (1959; 2nd ed., New York: Atheneum, 1973).
108. Dunn, "Prospects for Genetic Improvement," 32 (see n. 106).
109. Dunn, *Heredity and Evolution*, 109–22 (see n. 107). The sixth chapter (106–31) is titled "Isolated Populations and Small Communities." The five studies associated with the institute were Jewish Romans, Indian castes, Charleston's Gullah tribe, the James Island African Americans, and Black Caribs of British Honduras. Bentley Glass of Johns Hopkins University conducted the other study on Dunkers. Dunkers are a group descended from German Baptists who baptize through a triple immersion. They were originally called Tunkers and Tumblers, which meant "to dip," and at some point the name was inadvertently altered. *Oxford English Dictionary Online*, s.v. "Dunker, Tunker," http://dictionary.oed.com (accessed May 26, 2009).
110. L. C. Dunn and Stephen Dunn, "Jewish Community of Rome," 118–24 (see n. 5); and L. C. Dunn, *Heredity and Evolution*, 113–14 (see n. 107).

111. L. C. Dunn, "Preface to Revised Edition," in *Heredity and Evolution*, 2nd ed., v (see n. 107).
112. Bentley Glass, "Heredity and Evolution in Human Populations," *Quarterly Review of Biology* 34 (September 1959): 239–40; and L. C. Dunn, *Heredity and Evolution*, 130–31.
113. Hugh H. Smythe, "Race, Science, and Society," *Contemporary Sociology* 6 (January 1977): 77.
114. L. C. Dunn to William S. Pollitzer, 24 June 1960, Dunn Papers, series 1, box 24, William S. Pollitzer, 1960–62, American Philosophical Society, Philadelphia; L. C. Dunn to Leon Poliakov, 24 October 1961, Dunn Papers, series 1, box 24, Leon Poliakov, 1961, American Philosophical Society, Philadelphia.
115. L. C. Dunn to Landauer, Easter Sunday 1970, and L. C. Dunn to Landauer, 9 February 1971, Dunn Papers, series 1, box 20, Walter Landauer, 1970 and 1971, American Philosophical Society, Philadelphia; and Ethel Dunn, "Stephen P. Dunn," 48 (see n. 25). This post was one of the lengthier ones that Stephen held, from 1970 to 1974.
116. Louise Dunn to Theodosius Dobzhansky, 29 August 1974, Dobzhansky Papers, B: D65, Louise Dunn, American Philosophical Society, Philadelphia.
117. L. C. Dunn, *Heredity and Evolution*, 113–14 (see n. 107).
118. L. C. Dunn, *Heredity and Evolution*, 121 (see n. 107); and L. C. Dunn, "Lecture II: The Community as Viewed by a Biologist," in "Synopsis of Three Lectures," 18–23 (see n. 106).
119. L. C. Dunn, "Cross Currents," 11 (see n. 106).

Changes in Scientific Opinion on Race Mixing: The Impact of the Modern Synthesis

Paul Farber

In the early 1950s, the United Nations Educational, Scientific and Cultural Organization (UNESCO) issued two statements on race. They were crafted in response to a resolution passed by the organization at its fourth General Conference in 1949 as part of the organization's campaign against racism. The director-general had given the Social Science Department a charge to gather scientific information on the problems of race, to disseminate it, and to prepare an educational campaign.[1] Arthur Ramos, the head of the department, put together a group of social scientists. Although he died shortly thereafter, the project went ahead, and a committee began meeting in Paris toward the end of 1949. Before publication, a draft of the statement was circulated among some life scientists such as Theodosius Dobzhansky, L. C. Dunn, Julian Huxley, and Hermann Muller, as well as some social scientists such as Gunnar Myrdal. The final draft was written by Ashley Montagu. It soon drew criticism, partly because of the composition of the committee (social scientists), so a new committee (with L. C. Dunn as rapporteur) was appointed.[2] This committee consisted of geneticists and physical anthropologists, and it drafted the second statement.[3]

The UNESCO statements reputed the virulent scientific racism that had characterized the Third Reich and rejected the pervasive scientific racism that had developed for well over a century in most Western countries.[4] In part, the statements did this by defining human races in a manner that reflected the evolutionary biology of the time. The first statement defined the scientific use of the term *race* as populations of *Homo sapiens* that vary genetically from other populations. These populations "are capable of interbreeding with one another but, by virtue of the isolating barriers which in the past kept them more or less separated, exhibit certain physical differences as a result of their somewhat different biological histories."[5] The differences are in "the frequency of one or more genes. Such genes, responsible for the hereditary differences between men, are always few when compared to the whole genetic constitution of man and to the vast number of genes common to all human beings regardless of the population to which they belong."[6] Or, put more succinctly, "the term 'race' designates a group or population characterized by some concentrations, relative as to frequency and distribution, of hereditary particles (genes) or physical characters, which appear, fluctuate, and often disappear in the course of time by reason of geographic and/or cultural isolation."[7] Equally important, the statements distinguished the *scientific* use of the term *race* from *everyday* uses that commonly referred to national, religious, linguistic, or cultural groups. These "national, religious, geographic, linguistic and cultural groups do not necessarily coincide with racial groups; and the cultural traits of such groups have no demonstrated genetic connection with racial traits."[8] The statement also clearly rejected the long-held notion that race mixing produced biologically inferior results. "With respect to race-mixture, the evidence point unequivocally to the fact that this has been going on from the earliest times . . . no convincing evidence has been adduced that race-mixture of itself produces biologically bad effects."[9]

Historians have often cited the UNESCO statements as central documents in the rejection of scientific racism and, in particular, the idea that race mixing was biologically unwise.[10] They have also correctly noted that the UNESCO statements had deep political and social roots. The memory of the Second World War was fresh, and racial tensions were an important part of the cultural landscape in many parts of the globe. Scholars have argued that although the rejection of scientific racism and antimiscegenation thought had a strong ideological foundation, there was little in the way of new, empirical evidence to justify the stance. The scientific community's rapid acceptance of these positions, therefore, has often been portrayed by historians as a result of humanitarian or ideological factors, rather than a scientific shift.[11]

Without diminishing the importance of the cultural dimension of the rejection of scientific racism, I would like to suggest that part of the story—the scientific roots of the change—has generally been overlooked or misunderstood. Science and its misuse had indeed been a large part of the problem of scientific racism by providing a strong intellectual foundation to legitimate racism. But science also played a role in the solution to the problem by removing that foundation. To understand the importance of science in the rejection of scientific racism, looking at the change in positions on the topic of race mixing in the United States is particularly instructive, both as a case study and as an issue that represented an important underlying and emotionally charged aspect of the broader topic.

Anti-race mixing sentiment was prevalent in the U.S. scientific community during the early twentieth century. The subject was one that came under the purview of the eugenics movement, which sought to enlighten the public about the implications of research in genetics and on human heredity, purportedly from an evolutionary perspective.[12]

The movement focused much of its attention on both the problem of feeblemindedness in White populations and the threat of immigration from "inferior" European countries, rather than on the issue of race mixing between "Negroes" and "Caucasians"—not because eugenicists approved of such alliances, but rather because a variety of social and legal forces already discouraged such mixing.[13]

When eugenicists did discuss race mixing of Blacks and Whites, they did so in negative terms. They were concerned about two main problems: The first was that mixing would result in a dilution of the American race (i.e., Anglo-Saxon). The anthropological literature of the period ranked races on the basis of what anthropologists took to be the "accomplishments" of the races as reflected in history.[14] This literature portrayed Anglo-Saxons as among the highest, southern and eastern Europeans lower, the Irish below them, and Africans even lower. Race mixing between Whites and Blacks, the authors claimed, could only lower the potential of future generations, and therefore should be avoided.[15] Anglo-Saxon crosses with Negroes in this literature were considered to be worse than mixes with Slavs or Sicilians, since those of African descent were considered lower on the scale in human accomplishments. As distinguished geneticist Edward East wrote, "In reality the Negro is inferior to the white. This is not hypothesis or supposition; it is a crude statement of actual fact. The Negro has given the world no original contribution of high merit. By his own initiative in his original habitat, he has never risen. Transplanted to a new environment, as in the case of Haiti, he has done no better. In competition with the white race, he has failed to approach its standard."[16]

The second problem with race mixing, according to the authors of eugenic literature, was the concern that alliances of individuals from distant groups (e.g., Anglo-Saxon and African) ran the risk of producing "disharmonious crosses," an idea that had been popularized by Norwegian eugenicist Jon Mjoen through his studies of the offspring of Norwegians and Lapps in northern Norway. Mjoen claimed that these hybrids showed more disharmonious traits than the parent stocks. Some were minor (disproportionate extremities or large ears), but others were serious (higher rates of diabetes or lower resistance to tuberculosis). He believed that the higher rates of diabetes in Lapp-Norwegian hybrids reflected "glandular lesions" resulting from hereditary disharmonies.[17] Medical writers supported the view that crosses involving individuals from widely different races might give rise to "chaotic constitutions," and one review article in 1933 concluded, "It may be said that the bulk of medical opinion is against hybridization between the Primary Races and that the best eugenic opinion is definitely against it."[18]

The study of plant and animal breeding reinforced concerns about the biological effect of race mixing. Breeders had long worked to establish "pure lines" that had economic value and would breed true. East's widely read *Inbreeding and Outbreeding* argued that inbreeding could be used to isolate traits found in parental stocks. If rigorous selection was applied, a breed could be developed with highly desirable traits and potentially commercial value. According to East, crossbreeding represented an even better and more widespread method of improving stocks. "Hybrid vigor," or "heterosis," he noted, had given rise to impressive gains in horticulture and animal husbandry. One might think that East's enthusiasm for "hybridization" in corn and hogs might have inclined him to consider race mixing in humans as potentially positive, and he did to some extent. East devoted two chapters of his book to the consideration of inbreeding and outbreeding in humans. He concluded that some (but not all) crossbreeding of closely allied races has had relatively beneficial effects. The English and Scotch, for example, were the products of successive mixtures of closely related peoples and consequently were a highly variable population that produced genius (although also some "wretchedness"). The Irish, by contrast, were the product of close interbreeding of two "savage" tribes and consequently the "Irish have hardly a single individual meriting a rank among the great names of history, or a contribution to literature, art, or science of first magnitude."[19] The admixture of inferior stock (such as the Irish) to a superior stock (such as the English), according to East, led to a lowering of the superior one. East, consequently, had doubts about the wisdom of the United States' relatively open-door policy for immigration from European countries. A melting pot, according to him, must "be sound at the beginning, for one does not improve the amalgam by putting in dross."[20] What about crosses among races he considered further apart? Here East saw even greater potential problems, for such crosses could lead to a "breaking down of the inherent characteristics of each."[21] Although some positive recombinations were possible, they were unlikely because human reproduction was not guided (as was done in agriculture). The fruit of generations of selection and adaptation could be squandered if unwise crossbreeding was allowed to happen.[22]

East reflected the attitude of many geneticists in the early decades of the twentieth century who were influenced by the new Mendelian genetics that were uncovering the complexity of inheritance. From experimental work, for example, East knew that important traits could be the result of multiple genetic "factors" and that these combinations of factors could be enhanced by intensive breeding. More importantly for the issue of race mixing, East believed that naturally occurring races contained complex sets of compatible physical and mental traits. Human race mixing that involved individuals of "distant" groups posed a danger because the positively linked traits that had been selected over long periods of time might separate in the offspring of a mixed cross and might result in a loss of genetic improvement.

On theoretical grounds, interracial crosses did not appear to geneticists to be a wise social policy. Although East claimed that the natural inferiority of Negroes doomed them to extinction in the United States unless saved by "amalgamation," he concluded that "it seems an unnecessary accompaniment to humane treatment, an illogical extension of altruism, however, to seek to elevate the black race at the cost of lowering the white."[23]

The weight of scientific opinion was against race mixing, but this is not to say that the arguments were strong. What empirical evidence supported the claims? The most extensive and widely quoted study was the one undertaken in Jamaica by the geneticist and eugenics advocate Charles Davenport.[24] In 1926 Davenport and Morris Steggerda (then still a student) sailed to the racially mixed island of Jamaica on a research trip sponsored by the Carnegie Institution (with funding from the eugenics enthusiast Wickliffe Preston Draper[25]). The Jamaica study compiled hundreds of physical measurements and psychological test results on roughly three hundred adults from three populations: pure White, pure Negro, and hybrid (mulatto). (A larger number of children were also observed, but they played a small role in the work.) Although Davenport could point to some anthropometric differences among his groups, he had little evidence of any physical dangers from race mixing. His main finding was that a few hybrids had long legs (Negro trait) and short arms (White trait). In his official Carnegie Institution report in 1929, he confessed that "we do not know whether the disharmony of long legs and short arms is a disadvantageous one for the individuals under consideration."[26]

But if the empirical evidence was not robust, it didn't diminish confidence in the conclusion that race mixing was not biologically wise. Samuel J. Holmes of the University of California at Berkeley did an extensive survey of the literature on the subject. He admitted there was a lack of solid evidence but nonetheless concluded in 1936 that race mixing constituted a "dangerous experiment."[27]

By the 1940s, however, a significant reversal of opinion in the scientific community on the alleged biological consequences of miscegenation had taken place. One of the chief forces in this change was the anthropologist Franz Boas, who strongly disagreed with Davenport and argued that so much mixing had gone on in the past that no pure races existed. In his American Association for the Advancement Science Presidential Address, Boas noted that little evidence existed on the question of the biological effects of race mixing of "distant races" but nonetheless summarized his position in the following words: "I believe the present state of our knowledge justifies us in saying, that while individuals differ, biological differences between races are small. There is no reason to believe that one race is by nature so much more intelligent, endowed with great will power, or emotionally more stable than another, that the difference would materially influence its culture. Nor is there any good reason to believe that the differences between races are so great, that the descendants of mixed marriages would be inferior to their parents. Biologically there is neither good reason to object to fairly close inbreeding in healthy groups, nor to intermingling of the principle ones."[28] Boas and his students, particularly Otto Klineberg and Montagu, campaigned vigorously against the scientific racism that had characterized much of anthropology and genetics in the 1920s and 1930s.[29] They argued for a more relativistic approach to race that stressed environmental as well as hereditary factors.

The role of the "Boas School" has generally been recognized in the rejection of scientific racism and in opposing "anti-miscegenation" sentiment.[30] What has been not so widely recognized was the importance of the work being done in population genetics and evolution that was used by anthropologists like Montagu to bolster their positions.

Starting in the late 1930s, a set of theoretical works on the theory of evolution transformed biology. The Modern Synthesis, as it has come to be called, reestablished a Darwinian perspective on evolution and constructed a solid foundation in population genetics to build on. When Darwin published *The Origin* in 1859, he argued that contemporary life on Earth had descended from earlier forms and that they had changed in time. The theory explained many of the leading scientific questions of the day and synthesized what was known of natural history. Darwin's theory emphasized natural selection as the major cause of evolutionary change. As historians have demonstrated, Darwin was able to convince the scientific community, and later a wider public, that evolution had occurred on the planet, but he was not successful in establishing natural selection as the principle cause of those changes. By the end of the century, many in the scientific community, while accepting the general notion of evolution, considered "Darwinism" dead; that is, they thought that Darwin's reliance on natural selection as the main driving force in evolution was not justified. A number of alternative evolutionary theories were proposed, but none were generally accepted by the naturalist community. However, in the 1930s, work done in genetics, systematics, statistics, field biology, and a number of other areas was brought together into a new synthetic theory that had natural selection at its core. This new theory, the Modern Synthesis, also emphasized race in its explanation of how species come to evolve. A race (sometimes called a subspecies or a variety) was a population or set of populations that differed from other populations of the same species. Interbreeding kept all the races part of the same species, but if the individuals of a race came to be isolated in any of a number of fashions, they could begin to differentiate into a more and more different population. In time, the individuals of this race might become so different that they could no longer successfully breed with individuals of the parent race, and this race would therefore have become a new species. Biologists used the term *race* to refer to populations of any species that differed from other populations of the same species. The term was in no way restricted to groups of humans but was employed routinely to describe birds, insects, and plants. Indeed, the term predates evolutionary theory and had been used for a couple of centuries as a basic term of classification to refer to a subgroup of a particular species.

For the architects of the modern theory of evolution, race was a central concept, and consequently the theory held potential implications for evaluating the biological aspect of race mixing in humans. We can see this clearly in the work of Dobzhansky, who in 1937 published the first monograph setting out the newly emerging theory of evolution, *Genetics and the Origin of Species*.[31] Dobzhansky began his career working on ladybird beetles and developed an interest in the variations within populations of the same species. His scientific goals, even during his early career, transcended the desire to sort out technical problems in insect classification. Instead, he sought to understand the process of evolution and its implications for humans. After coming from the Soviet Union to the United States in 1927 to work in Thomas Hunt Morgan's laboratory, Dobzhansky focused his studies on the genetic differences in populations of *Drosophila pseudoobscura*, a fruit fly that inhabits the southern and western parts of the United States and extends down into Mexico.[32]

Dobzhansky found this species fascinating because of the work of Donald Lancefield, a former student of Morgan's who in 1929 described two "races" of the species. They were of interest because although the "races" appeared morphologically the same, when individuals from the different "races" were crossed they produced sterile males but fertile females. Luckily, Dobzhansky had a student from the Seattle area where both races of *D. pseudoobscura* could be found, and his student collected them for him during the summer of 1932. Thus began a long experimental program with *D. pseudoobscura*, with Dobzhanksy himself collecting samples for years.

Historians like William Provine have described how central this species was for Dobzhansky's pioneering work in evolution, for it led him to support the "biological species concept" that defines species as sets of interbreeding populations that are reproductively isolated from other sets of populations. The definition stresses the dynamic, rather than static, nature of species.[33] For numerous reasons and in a variety of ways, a subpopulation can begin to develop "isolating mechanisms," which in time can lead to its reproductive isolation from the parent populations; that is, a subpopulation (i.e., a race) can become a new species. Ultimately, Dobzhansky argued that because of their reproductive isolation the two "races" described by Lancefield (which, in gross anatomical traits, looked identical) are actually separate species. The distinction that reproductively isolated races are different species was critical and held significant implications.

Race, according to Dobzhansky, is a "tool for description not of individuals, but of subdivisions of species."[34] Races differ in their gene frequencies and arrangements. By looking at an individual's characteristics, one might be able to determine the *probability* of its race, but only the probability. The individual could be from any race. Dobzhansky stressed that races are *open* systems, whereas species are *closed* systems. By this he meant that since individuals of different species are not able to reproduce successfully, species do not regularly exchange genes.[35] Individuals, in this sense, belong to one species or another. In contrast, genes regularly flow from race to race since individuals are interfertile. An individual might carry some genes frequently found in one race but also carry a few genes most frequently found in a different race.[36] Since races are open systems, they can be defined in a number of different ways depending upon what characteristics are of interest to the researcher. Subpopulations can be of varying geographical size, for example.

When applied to *Homo sapiens*, these distinctions suggested a new conception of the biological significance of human race. Dobzhansky was always quick to note that our everyday conceptions of race were wholly inadequate as biological concepts and our knowledge of human genetics primitive. The lack of consensus in physical anthropology over the classification of human races reflected the difficulty of adequately defining them. Nonetheless, Dobzhansky believed that human populations had experienced sufficient geographic isolation to create populations with meaningfully different frequencies of certain genes. The vast amount of mixing that has also occurred has maintained the status of races as populations of the same species. And since genes can vary independently, by focusing on a particular gene or small set of genes (or more accurately, their expression), one can validly classify humans in quite a number of ways. Skin color doesn't correlate, for example, with blood type. A classification based on blood type would be different than one based on skin color or eye color. Hence there is a certain arbitrariness in classifications.

What about the disharmonies that earlier geneticists had worried would result from race mixing? Dobzhanksy held that races reflect a stage in evolution: a population for any of a number of reasons (e.g., adaptive pressure, genetic drift) comes to have a discernibly different set of gene frequencies than other populations. The greater the differences in gene frequencies and the more the population is geographically isolated, the greater the chance that the race will develop complexes of genes adapted to a specific environment and begin to develop isolating mechanisms that restrict gene flow. So, theoretically, races of humans that contained individuals with adaptive complexes might exist. But when individuals of different races crossbreed, their offspring might possess genes that in their parents belonged to adaptive complexes but in the offspring had become disaggregated and recombined into less-fit combinations. Dobzhansky argued that although gene complexes could favor survival and reproduction (i.e., be adaptive) and hybridization could, theoretically, lead to less well-adapted combinations, scientists had no empirical evidence of such breakdowns in fitness. Fear of disharmonies was simply "farfetched."[37] Different human populations had mixed extensively. Dobzhansky used many examples to illustrate his point that this mixture had not resulted in any patterns of disharmonious constitutions.

One might inquire why Dobzhansky had confidence that race mixing posed no serious biological problem, whereas Davenport and many others who had established reputations in genetics had stated otherwise. Here, Dobzhansky's perspective, built from years of breeding *Drosophila*, was important. A race for Dobzhansky was an open system with substantial variation. He had found that the frequencies of particular genes varied in subpopulations over seasons and over relatively small geographic locations. Mixing was constantly occurring and was critical for keeping the species intact.

The process of evolution, of course, can lead to significant divergence and, with that, the creation of isolating mechanisms. Although Dobzhansky recognized many forms of isolating mechanisms, what they all had in common was that they prevented crosses or caused crosses with individuals from different groups to have either no offspring or less fertile offspring. When it came to humans, there was no credible literature that could substantiate the claim that individuals from different populations exhibited the slightest level of sterility. Many had speculated on the difficulty of successful mixing, to be sure, and it had become a common theme in nineteenth-century literature on Jews, the Irish, and Negroes.[38] To Dobzhansky, this literature was highly suspect, not only because of its stereotypes and speculative nature, but also because it wasn't clear that the authors of these writings had adequately identified populations that constituted races. For example, Dobzhansky didn't believe the Jews constituted a race, and he believed that the "Negro" in its American context was hopelessly too broad a term to be meaningful.[39] At issue, of course, was the method commonly used in anthropology to define different "races"—a methodology that the Modern Synthesis firmly rejected. Physical anthropologists and earlier geneticists had categorized groups by tabulating the mean values of a set of physical measurements. Davenport relied on an elaborate set of measurements that he used in comparing different populations.[40] He argued that the mean values of these measurements differed among races, and in describing the potential dangers of race mixing, he contrasted the mean sizes of various traits in different groups. Dobzhansky argued that comparing means had no value, for populations consisted of individuals while mean values were the properties of groups. The variation and complexity of the possible combinations of genes, moreover, was so great that comparing the mean values of a few selected traits proved a meaningless exercise. Humans shared so many genes and the variation within a population was so large that specific individuals from different populations might be more alike than individuals from the same population. Concentrating on mean values suggested some ideal "type," a notion Dobzhansky and his colleagues such as Ernst Mayr were trying to drum out of existence in biology.

Dobzhansky's genetics and evolutionary views undercut opposition to race mixing in a number of important ways. First, the objection to race mixing generally involved a confusion of categories. Races are populations. Races mix, not individuals. Races have a normal degree of gene exchange with other races. If they don't, a process of divergence sets in, and the races become different species. For those races that are part of the same species, as is the case with humans, a normal degree of exchange is expected. Moreover, all the races that exist today are the products of extensive mixing of individuals in the past. There are no "pure" races. According to Dobzhansky, even the most extensive attempt to separate populations, the Indian caste system, showed no evidence of having produced genetically "pure" groups.[41]

Second, when discussing race mixing, earlier scientists had compared mean values of traits. This made no sense to Dobzhansky; means don't mix, individuals in populations do, and the variation within any population is enormous. There are "Negroes" who are whiter than some "Caucasians" because of the wide range of variation within each race. Moreover, traits are independent, so even by looking at means, we would be in a hopeless muddle when it comes to defining the popularly perceived races; human traits such as blood type, skin color, and hair type are not closely correlated.

A third issue had to do with the empirical record. Whites mix with Blacks, and there is no evidence of sterility in the offspring or "disharmonious" constitutions resulting. Also, there is no evidence that the occasional poor combination of genes, mostly deleterious recessives, is more frequent in mixed crosses than in crosses within a race.

For Dobzhansky, then, earlier attempts to discourage race mixing based on alleged scientific arguments were invalid. He made efforts to popularize his position, and he was prominent in various symposia and edited volumes that addressed the issue.[42] Through the work of physical anthropologists like Montagu and later Sherwood Washburn, Dobzhansky's perspective based on population genetics had a huge influence on how race was discussed. Montagu was perhaps the most aggressive in popularizing the implications of Dobzhansky's genetics for the understanding of race and race mixing. As early as April 1940, Montagu presented a paper at the annual meeting of the American Association of Physical Anthropologists in which he laid out how the "new" genetics altered the meaning of race as used by anthropologists—that is, their attempts to define a set of metrics that characterized the different races. *Race* for Montagu had a new biological meaning. The human species consisted of a large set of populations that were the descendants of some ancestral population. In time, the ancestral population had dispersed and had become geographically separated into multiple populations. These populations were subject to random variations and have come to vary in their gene frequencies. Moreover, the individuals in these populations were also subject to gene mutations at different rates and in different characteristics. Montagu stressed the dynamic aspect of race: populations were in constant flux, and what we call a *race* was really just a population with a particular set of gene frequencies. These frequencies were shifting, and they were operated on by secondary forces such as migration, social and sexual selection, endogamy, and exogamy.[43] There were, then, two sets of forces molding human populations: geographical variability in local groups and the social forces of "mixing."

According to Montagu, the then-current definitions of race were woefully inadequate because they focused upon "arbitrary and superficial selection of external characters."[44] Quoting Dobzhansky, Montagu noted that the geographical distribution of genes that were allegedly responsible for these physical differences were frequently independent: blood group distribution did not map onto skin color distribution. Montagu went beyond Dobzhansky and concluded with a proposed new terminology. Since *race* represented such a complex and confusing concept, he chose to use the term *ethnic group*, which Huxley and A. C. Haddon used in their widely read book, *We Europeans: A Survey of "Racial" Problems*, published in 1936.

Huxley, the prime author of *We Europeans*, argued against the Aryan racial theories emerging from Hitler's Germany and stressed the inherently mixed nature of European "races." The biological concept of race as a variety of subspecies, according to Huxley, could not be applied to groups such as the "Germans" and the "French," so he suggested using the new term *ethnic group*, which he believed held fewer negative connotations. Huxley didn't confine his treatment just to European groups. All the groups recognized by anthropologists as races were mixtures of peoples, he claimed, so no "pure" races existed anywhere at present.[45] Consequently, the social policy that forbid race mixing "turns out not to be primarily a matter of 'race' at all, but a matter of nationality, class or social status."[46] Huxley concluded that "racialism is a myth, and a dangerous myth at that. It is a cloak for selfish economic aims which in their uncloaked nakedness would look ugly enough. And it is not scientifically grounded."[47]

Montagu agreed with many of Huxley's views: the use of the term *race*, his suggestion that *ethnic group* replace the former concept, and the fundamental genetic unity of humans. For Montagu, ethnic groups are merely populations of *Homo sapiens* that have different relative frequencies of some genes, and these differences are maintained (or not maintained) by physical and social barriers. If Negroes and Whites were "freely permitted to marry, the physical differences between Negroes and Whites would eventually be completely eliminated through the more of less equal scattering of their genes though the population."[48]

In the first edition of his most well-known book, *Man's Most Dangerous Myth: The Fallacy of Race*, published in 1942, Montagu brought together (with some rewriting and revision) a number of articles he had published on race. He extended Huxley's warning that racialism was a dangerous myth to the broader claim that the concept of race itself was a myth. Montagu's book went through six editions and reflected an ever-more strident stand on the issue. In the first edition, Montagu admitted that there was a biological meaning of race: "mankind is comprised of many groups which are often physically sufficiently distinguishable from one another to justify their being classified as separate races."[49] These are very large groups: Mongolian,

Caucasian, Negro, Australo-Melanesian, and Polynesian. "Races," he reminded us, are temporary mixtures that are in flux and, as experience has shown, among which there is complete interfertility. Montagu argued that the genetic differences among these groups are slight; that is, they differ in the distribution of a small number of genes. More importantly, he went on to state that the many complex groups to which we normally apply the term *race* are the products of cultural as well as physical difference, and that the mixture among these groups is considerable. In this sense, *race* has lost its anthropological and biological significance.

Race mixing was a major topic in Montagu's book, which had a long chapter that explored many dimensions of the issue. Montagu made it clear that objections to race mixing were basically reflections of social factors, not biological ones, and that "there can be little doubt that those who deliver themselves of unfavorable judgments concerning race-crossing are merely expressing their prejudices."[50] Taking the offensive, he moved the discussion to the general issue of hybridization. Montagu briefly reviewed the importance of crossbreeding and outbreeding in agriculture and then pointed out the irony of attempting to argue against race mixing with a reference to hybridization research:

> *It is, indeed, a sad commentary upon the present condition of western man that when it is a matter of supporting his prejudices he will so distort the facts concerning hybridization as to cause laws to be instituted making it an offense against the state. But when it comes to making a financial profit out of the scientifically established facts, he will employ geneticists to discover the best means of producing hybrid vigor in order to increase the yield of some commercially exploitable plant or animal product. But should such a geneticist translate his scientific knowledge to the increase of his own happiness and the well-being of his future offspring, by marrying a woman of another color or ethnic group, the probability is that he will be promptly discharged by his employer.*[51]

Montagu most likely had in mind the Racial Hygiene laws in Germany when he wrote this book in 1942, but he certainly also intended the critique to apply to U.S. antimiscegenation laws and to earlier geneticists like East who stressed the value of outbreeding in hogs and corn but cautioned against it in humans.

Montagu also provided a review of the existing studies that focused on race mixing and their interpretations. One of the points he attempted to establish was how much the social place of crossbreeds affected their well-being and achievement. The issue was not heredity but culture. In those places where the offspring of a mixed race couple were not discriminated against, they did as well as nonmixed offspring. Physically there was even the suggestion of hybrid vigor in some cases. Montagu ridiculed Davenport and Steggerda's figures, showing an alleged disharmony in leg and arm length in Jamaican Browns, and stated unequivocally, "The whole notion of disharmony as a result of ethnic crossing is a pure myth. . . . The differences between human groups are not extreme enough to be capable of producing any disharmonies whatever."[52]

Montagu concluded the first edition of *Man's Most Dangerous Myth* with an appendix titled "State Legislation Against Mixed Marriages in the United States." The data was from Chester Vernier's *American Family Laws*,[53] and Montagu quoted Vernier's remarks on the data: "Such legislation is not based primarily upon physiological, psychological, or other scientific bases, but is for the most part the product of local prejudice and of local effort to protect the social and economic standards of the white race."[54] Montagu, a man often ahead of his time, also noted that the antimiscegenation laws contravened the United States Constitution, but that the Supreme Court had never handed down a decision that related to them.

Subsequent editions of *Man's Most Dangerous Myth* expanded the basic arguments about race and race mixing and grew from the first edition's 214 pages to 699 in the 1997 sixth, and final, edition.[55]

Dobzhansky appreciated Montagu's forceful statements on race and race mixing. He was not, however, willing to abandon the term *race*. Montagu argued that the term had been abused to such a degree that the cultural baggage it carried was so toxic and that its usage had varied so greatly that, as a consequence, its biological meaning had been compromised. But major concepts in science, although often radically modified (e.g., that of species) are not easily dropped, and it should come as no surprise that Dobzhansky didn't see the value in eliminating a term that had been central to his research. Although Dobzhansky accepted the view that current, everyday racial categories were social constructions and that races could be classified in many ways—the principal criterion being convenience—he nonetheless held that *race* had biological meaning; that is, humans group into Mendelian populations (groups of interbreeding individuals). These populations are mobile and in constant flux, but they are nonetheless "real."

The argument was not limited to semantics. Those who preferred the term *ethnic group* considered such designations as largely social constructs. They were reacting to the bogus racial theories of the Nazis that depicted Jews as a separate race and divided Europeans into sets of different races. In many ways, the issue had to do with what appeared to be confusion in the meanings of *nationality*, *ethnicity*, and *race*. As much as Dobzhansky abhorred the abuse of the term *race* as applied to humans, and as much as he realized the arbitrary nature of any particular classification of human races, he remained tied to his population biology perspective that recognized species as consisting of many interbreeding populations (i.e., races). To agree that races were purely social constructs was to concede the field to those who misused science, and, moreover, he didn't believe it would result in any reduction of discrimination. To someone like Dobzhansky, who was painfully aware of the dangers of letting social priorities dictate to science (the research school of population genetics he came from was under attack in the U.S.S.R. and would ultimately be abolished), such a move was unacceptable and unwise.[56] Ideology had deformed science. Nazi "racial science" had shown the dangers of putting social policy before rigorous science, and political attacks on Mendelian genetics in the Soviet Union demonstrated the damage that could be done to science when ideology overcame scientific evaluations. Scientific racism in the United States had deformed the life sciences for many years and served to justify slavery, limit immigration, and outlaw race mixing. The answer, according to Dobzhansky, was not to give in to the misuse of science but to assert the value of scientific investigation and guard its independence.

Dobzhansky and Montagu sparred over this issue for years. In a letter to Montagu in 1944, Dobzhansky wrote:

> *It is hardly necessary to say that I admire your book, and regard its function as contributing to sanity as important as any book can make itself to be. My criticisms, at least most of them, are due to not only academic fuss over details, but, and principally to fear of what use these details can be made of by the opposition. It is obvious I think that the racialists which were so overwhelmingly strong in USA before 1932 and before Hitler has made Madison Grant's view unpopular, have not disappeared since. They are just laying low and waiting for their day, which may come as soon as the war ends or even sooner. Just think what will then be written by the eugenists—every little slip of yours will be used to show that all that you wrote is wrong.*
>
> *It is for this reason that I can not subscribe to your campaign against the existence of races and for the ethnic groups or divisions or whatever you want to call them. The only way is to divest the word race of it [sic] emotional contents; and if we biologist can help in this, we shall justify our existence. Surely biologists by themselves can not do it 100%! The propagandist trick of making people swallow something under a different name might be useful in salesmanship and politics but I am afraid of its consequences in science.*[57]

Montagu did not accept Dobzhansky's point of view and continued to argue against the use of *race* in anthropology. Although the two were personal friends and supported one another in a number of projects, they never came to agreement on this matter. In 1961 Dobzhansky, somewhat lightly, wrote to Montagu about the latter's autobiography: "The chapters on 'Ethnic group and race' is, of course, deplorable. But let us say that it is good that in a democratic country any opinion, no matter how deplorable, can be published."[58] Neither man had the last word on the subject. Montagu did not succeed in removing the term *race* from anthropology, but he was successful in helping convince the anthropological community that the term had no biological significance and referred to cultural groups.[59]

The dispute with Montagu over terminology reflected Dobzhansky's concern with the social use of science. Among the architects of the Modern Synthesis in the United States, Dobzhansky was the most outspoken on social issues. In addition to the symposia he organized and the articles he wrote, in 1946 he collaborated with Dunn on *Heredity, Race, and Society*, a book written for the general public on the implications of modern genetics for social issues.[60]

We can see quite clearly Dobzhansky's influence in the UNESCO statements. The definition of race as consisting of subpopulations that shift in time and are constantly mixing came straight out of the Modern Synthesis and was built on Dobzhansky's population genetics. Indeed, some of the language describing race is identical to Montagu's contribution on race to a symposium organized by Dobzhansky and Washburn in June 1950 and explicitly cites Dobzhansky.[61] The rejection of scientific racism and of antimiscegenation sentiments followed directly from the new scientific foundation for thinking about race and race mixing. The UNESCO statements were clear manifestations of the shift in thinking, and since they were part of an international educational campaign, they had considerable importance.

As stated earlier, the political events leading up to and surrounding the UNESCO statements of the early 1950s played an important role in their coming into existence and in their content. UNESCO was involved in a global educational campaign to reduce racism, counter antimiscegenation sentiment, and promote the ideas of universal brotherhood. The UNESCO statements, however, also reflected a significant shift in the biology of race. In particular, the biology of the Modern Synthesis, which had reconceptualized the notion of race from a physical "type" to a subpopulation of a species, totally undercut previous biological justifications for opposing race mixing. The idea that race mixing was a "dangerous experiment" had been rejected by scientists like Dobzhansky and no longer had broad scientific acceptance.

In light of historians' legitimate censure of life scientists for their support before the 1940s of dubious eugenic ideas regarding race and race mixing, it is useful to consider the role of science in rejecting those ideas. We can learn a lesson from considering Dobzhansky and the impact of ideas like his on the history of scientific judgments about race mixing—that is, that science can (and should) be used to clarify our picture of nature and unmask false attempts to legitimate social prejudice. There is ample evidence from the past that science has been used to legitimate exploitive practices and justify discrimination. And the scientific community has rightly been taken to task for it. Scientific racism is a shameful blot on the integrity of the life sciences. But in telling that story, we need to be mindful that science was part of the solution. As such, the evolutionary biology associated with the Modern Synthesis should be considered, along with social and cultural forces, as among the important factors in the rejection of racism and antimiscegenation sentiment.

Notes

1. Ashley Montagu, *Statement on Race; An Annotated Elaboration and Exposition of the Four Statements on Race Issued by the United Nations Educational, Scientific, and Cultural Organization* (New York: Oxford University Press, 1972), 1.
2. For example, the *American Journal of Physical Anthropology* 6, no. 1 (1951): 3 had an editorial stating that professional anthropologists were astonished to be "by-passed" in the selection of the UNESCO panel.
3. Ibid., 5. For more on the statement, and subsequent ones by UNESCO, see Elazar Barkan, "The Politics of the Science of Race: Ashley Montagu and UNESCO's Anti-racist Declarations," in *Race and Other Misadventures: Essays in Honor of Ashley Montagu on His Ninetieth Year*, ed. Larry Reynolds and Leonard Liebermann, 96–105 (Dix Hills, NY: General Hall, 1996). Also see the interesting article by Michelle Brattain, "Race, Racism, and Anti-Racism: UNESCO and the Politics of Presenting Science to the Postwar Public," *American Historical Review* 112, no. 5 (2007): 1386–1413.
4. An extensive literature on the history of scientific racism exists. See, for example, Elazar Barkan, *The Retreat of Scientific Racism: Changing Concepts of Race in Britain and the United States between the World Wars* (Cambridge: Cambridge University Press, 1992); Mark Haller, *Eugenics: Hereditarian Attitudes in American Thought* (New Brunswick, NJ: Rutgers University Press, 1984); Nancy Stepan, *The Idea of Race in Science: Great Britain, 1800–1960* (Hamden, CT: Archon Books, 1982); Pat Shipman, *The Evolution of Racism: Human Differences and the Use and Abuse of Science* (New York: Simon and Schuster, 1994); Audrey Smedley, *Race in North America: Origin and Evolution of a Worldview* (Boulder, CO: Westview Press, 1993); and Sandra Harding, ed., *The "Racial" Economy of Science: Toward a Democratic Future* (Bloomington: Indiana University Press, 1993).

5. Montagu, *Statement on Race* (see n. 1).
6. Ibid.
7. Ibid., 8.
8. Ibid.
9. Ibid., 10.
10. Ibid., 502–6. The second statement closely echoed the first in its definition of race and its opinion on race mixing. The second statement goes into slightly more detail. The main difference between the statements is the omission of the statement claiming that biological studies support an ethic of universal brotherhood.
11. See William Provine, "Geneticists and the Biology of Race Crossing," *Science* 182 (1973): 790–96; Barkan, *Retreat of Scientific Racism* (see n. 4); and Joseph Graves, *The Emperor's New Clothes: Biological Theories of Race at the Millennium* (New Brunswick, NJ: Rutgers University Press, 2001).
12. Many studies of the eugenics movement exist. Useful background for this paper can be found in Garland Allen, "The Eugenics Record Office at Cold Spring Harbor, 1910–1940: An Essay in Institutional History," *Osiris*, 2nd ser., 2 (1986): 225–64; Diane Paul, *Controlling Human Heredity, 1865 to the Present* (Atlantic Highlands, NJ: Humanities Press, 1995); Daniel Kevles, *In the Name of Eugenics: Genetics and the Uses of Human Heredity* (Berkeley: University of California Press, 1985); Hamilton Cravens, *The Triumph of Evolution: American Scientists and the Heredity-Evolution Controversy, 1900–1941* (Philadelphia: University of Pennsylvania Press, 1978); Haller, *Eugenics* (see n. 4); Edward Larson, *Sex, Race, and Science: Eugenics in the Deep South* (Baltimore, MD: John Hopkins University Press, 1995); Kenneth Ludmerer, *Genetics and American Society: A Historical Appraisal* (Baltimore, MD: John Hopkins University Press, 1972); Lyndsay Farrall, *The Origins and Growth of the English Eugenics Movement, 1865–1925* (New York: Garland Publishing, 1985); Wendy Kline, *Building a Better Race: Gender, Sexuality, and Eugenics from the Turn of the Century to the Baby Boom* (Berkeley: University of California Press, 2001); and Mark Largent, "'The Greatest Curse of the Race': Eugenic Sterilization in Oregon, 1909–1964," *Oregon Historical Quarterly* 103, no. 2,(2002): 188–209.
13. For example, Larson, *Sex, Race, and Science* (see n. 12) states that the eugenics movement in the South was not especially concerned about Blacks or about "miscegenation," which was illegal. This is, of course, not to suggest extensive race mixing hadn't been taking place since early times among European settlers, Indians, and Africans (slaves or free). The vast majority of the literature on race mixing dealt with White/Black mixing. Other mixes were, of course, considered. This study concentrates primarily on the issue of White/Black mixing. In this essay, I have used the terms *White*, *Black*, *Negro*, *African*, and *Caucasian*, among others, as proper nouns that reflect the use of the terms at the time, not as modern scientific terms.
14. See John Haller, Jr., *Outcasts from Evolution: Scientific Attitudes of Racial Interiority, 1859–1900*, (Urbana: University of Illinois Press, 1971); and George W. Stocking, Jr., *Victorian Anthropology* (New York: Free Press, 1987) for the background of these rankings.
15. Madison Grant was among the most outspoken on the dangers of immigration. See his *The Passing of the Great Race* (New York: Charles Scribner, 1916). See

also Nathan Fasten, *Origin Through Evolution* (New York: Alfred Knopf, 1929). Although ranking of races was a popular notion, there was also a strong sense among the eugenicists that the bulk of immigrants from Europe were from the lowest strata of the countries from which they came. This made the situation even worse. See Paul Popenoe and Roswell Johnson, *Applied* Eugenics (New York: Macmillian, 1918).

16. Edward East and Donald Jones, *Inbreeding and Outbreeding: Their Genetic and Sociological Significance* (Philadelphia: Lippincott, 1919), 253.
17. See Jon Alfred Mjoen, "Harmonic and Disharmonic Racecrossings," *Scientific Papers of the Second International Congress of Eugenics* (Baltimore, MD: Williams and Wilkins, 1923), 41–61. This meeting took place at the American Museum of Natural History in New York, and consequently Mjoen's views became widely known in the United States. See also his "Biological Consequences of Race-Crossing," *Journal of Heredity* 17, no. 5 (1926): 175–85; and his "Race-Crossing and Glands: Some Human Hybrids and Their Parent Stocks," *Eugenics Review* 23 (1931): 31–40.
18. K. B. Aikman, "Race Mixture," *Eugenics Review* 25, no. 3 (1933): 161. The position was contested by the Harvard geneticist William Castle, but it is not clear that it had much effect. See Provine, "Geneticists and the Biology of Race Crossing," 792 (see n. 11). See also Paul, *Controlling Human Heredity*, 112–13 (see n. 12).
19. East and Jones, *Inbreeding and Outbreeding*, 257 (see n. 16).
20. Ibid., 264. This, of course, was the very opposite of the melting pot metaphor used by Israel Zangwill, who wrote the play *The Melting Pot* (New York: Macmillan, 1909) that introduced the term into popular vocabulary.
21. Ibid., 252.
22. This move allowed East and others to deal with Asians, who clearly had an impressive higher culture.
23. East and Jones, *Inbreeding and Outbreeding*, 254 (see n. 16).
24. On Davenport, see Mark Largent, *Breeding Contempt: The History of Coerced Sterilization in the United States* (New Brunswick, NJ: Rutgers University Press, 2007). A useful outline of Davenport's career and bibliography is Oscar Riddle, "Charles Davenport," *Biographical Memoirs, National Academy of Sciences of the United States of America* 23 (1945): 75–110. All the major works listed earlier dealing with eugenics, of course, spend considerable time on Davenport. In addition to Davenport's, a few other studies were routinely cited on the biology of race mixing. The most common were Eugen Fischer, who studied the Boer-Hottentot "hybrids" of South Africa, and Melville Herskovits, who studied the mulatto population in the United States. Fischer found no obvious biological problems, nor reduced fertility, in his group of subjects from mixed-race backgrounds. Fischer's work was problematic because he was not able to establish careful pedigrees for his subjects. Eugen Fischer, *Die Rehobother Bastards und das Bastardierungsproblem beim Menschen* (Jena, Germany: Gustav Fischer, 1931). Herskovits studied "the American Negro" and claimed that was an amalgam; that is, most individuals who identified themselves as "Negro" were not "pure" but a mixture of Negro, White, and a small amount of Indian. Herskovits claimed that a blending of physical racial traits were seen in his data (rather than a Mendelian segregation of characteristics), but did not explore the "superiority"

or "inferiority" of these traits. Melville Herskovits, *The American Negro* (New York: Alfred Knopf, 1928). It should also be noted that William Castle, one of Davenport's students and an important geneticist, strongly contested Mjoen's experiments and many of his conclusions.

25. Draper, who died in 1972, supported a number of eugenic initiatives. He was the main benefactor of the Pioneer Fund (incorporated in 1937), whose first president was Harry Laughlin, longtime associate of Davenport. Draper later supported a number of segregationist causes. See Michael G. Kenny, "Toward a Racial Abyss: Eugenics, Wickliffe Draper, and the Origins of the Pioneer Fund," *Journal of the History of the Behavioral Sciences* 38 (2002): 259–83; and Kenny, "A Question of Blood, Race, and Politics," *Journal of the History of Medicine* 61, no. 4 (2006) 456–91.
26. C. B. Davenport, Morris Steggerda, et al., *Race Crossing in Jamaica*, Carnegie Institution of Washington Publication no. 395 (Washington, DC: Carnegie Institution, 1929), 471.
27. Samuel J. Holmes, *Human Genetics and Its Social Import* (New York: McGraw Hill, 1936), 356.
28. Franz Boas, "Race and Progress," *Science* 74, no.1905 (1931): 6.
29. See Otto Klineberg, *Race Differences* (New York: Harper & Brothers, 1935) and the many editions of Ashley Montagu, *Man's Most Dangerous Myth: The Fallacy of Race* (1st ed., New York: Columbia University Press, 1942).
30. The term *miscegenation* came from a political hoax. The term was used as the title of an anonymous pamphlet and argued for its virtues. David Croly, the journalist who wrote the pamphlet, had hoped thereby to inject the issue of race mixing into the political campaign of 1864. He wished to make it appear as if the Republican Party supported race mixing. See Sidney Kaplan, "The Miscegenation Issue in the Election of 1864," *Journal of Negro History* 34, no. 3 (1949): 274–343. The term occasionally has a negative connotation, so I have refrained from using it except to indicate contemporary usage.
31. For a discussion of the origin of this book, see Joe Cain, "Co-opting Colleagues: Appropriating Dobzhansky's 1936 Lectures at Columbia," *Journal of the History of Biology* 35, no. 2 (2002): 207–219. For Dobzhansky's life and career, see Mark Adams, ed., *The Evolution of Theodosius Dobzhansky*, (Princeton, NJ: Princeton University Press, 1994); and E. B. Ford, "Theodosius Grigorievich Dobzhansky," *Biographical Memoirs of Fellows of the Royal Society* 23 (1977): 58–89. William Provine, *Sewell Wright and Evolutionary Biology* (Chicago: University of Chicago Press, 1986) has an excellent discussion of Dobzhansky's writings on genetics.
32. Various *Drosophila* species were used in genetic experiments. See Robert Kohler, *Lords of the Fly: Drosophila Genetics and Experimental Life* (Chicago, University of Chicago Press, 1994).
33. There were several versions of the biological species concept. The most famous was formulated by Ernst Mayr. For Dobzhansky, see William Provine, *Sewall Wright* (see n. 31).
34. Theodosius Dobzhansky and Carl Epling, *Contributions to the Genetics, Taxonomy, and Ecology of* Drosophila pseudoobscura *and Its Relatives*, Carnegie Institution of Washington Publication no. 554 (Washington, DC: Carnegie Institution, 1944), 138. See also Dobzhansky, "The Race Concept in Biology," *Scientific Monthly* 52, no. 2 (1941): 161–65.

35. Any hybrids that might be produced are either sterile or less fertile and therefore not "successful."
36. Dobzhansky and Epling, *Contributions to Genetics*, 49 (see n. 34).
37. Theodosius Dobzhansky, "The Genetic Nature of Differences Among Men," in *Evolutionary Thought in America*, ed. Stow Persons, 86–155 (New York: George Braziller, 1956).
38. A speculative literature existed that claimed mulattoes were infertile, or physically inferior, to Negroes or Whites, but by the late 1930s this literature had no scientific credibility. An often-cited text in this earlier literature on the problems of Negro/White "hybrids" is a translation of Paul Broca, *On the Phenomena of Hybridity in the Genus Homo* (London: Longman, Green, Longman and Roberts, 1864), which claimed that the genital organs of the different races (e.g., Negro male and White female) were of differing sizes and made certain crosses difficult or impossible.
39. See Dobzhansky, "Genetic Nature of Differences," (see n. 37)
40. See Charles Davenport, *Guide to Physical Anthropometry and Anthroposcopy* (Cold Spring Harbor, NY: Eugenics Research Association, 1927).
41. Dobzhansky, "Genetic Nature of Differences" (see n. 37). This argument was repeated in many other articles in which Dobzhansky discussed human evolution.
42. See, for example, *Origin and Evolution of Man*, Cold Spring Harbor Symposia on Quantitative Biology 15 (Cold Spring Harbor, NY: Biological Laboratory / Long Island Biological Association, 1951).
43. Ashley Montagu, "The Genetic Theory of Race and Anthropological Method," *American Anthropologist*, n.s., 44 (1942): 374.
44. Ibid.
45. Huxley thought that some "primary" races might have hypothetically existed in the distant past, but we have no direct knowledge of them.
46. Julian Huxley and A. C. Haddon, *We Europeans: A Survey of "Racial" Problems* (New York: Harper & Brothers, 1936), 232.
47. Ibid. A perhaps not surprising but ironic result of the reaction against Nazi racial science, which stressed the superiority of Nordic types, was the narrowing of the concept of race from a concept that applied to many groups we would today consider nationalities (e.g., French, Irish) to one that focused primarily on color. See Matthew Frye Jacobson, *Whiteness of a Different Color: European Immigrants and the Alchemy of Race* (Cambridge, MA: Harvard University Press, 1998).
48. Montagu, "Genetic Theory of Race," 375 (see n. 43).
49. Montagu, *Man's Most Dangerous Myth*, 4 (see n. 29).
50. Ibid., 99.
51. Ibid., 102–103.
52. Ibid., 119.
53. Davenport published an earlier tabulation of state laws. See Charles Davenport, *State Laws Limiting Marriage Selection Examined in the Light of Eugenics*, Eugenics Record Office, Bulletin no. 9, (Cold Spring Harbor, NY: Eugenics Record Office, 1913). For Vernier, see Chester Vernier, *American Family Laws: A Comparative Study of the Family Law of the Forty-eight States* (Stanford, CA: Stanford University Press, 1931).
54. Montagu, *Man's Most Dangerous Myth*, 188 (see n. 29).

55. An interesting research project would be to compare the various editions and show how they map onto the social issues of the day. The first edition has an appendix of miscegenation laws that does not appear later since the issue was moot after the 1967 Supreme Court decision *Loving v. Virginia.*
56. See Zhores Medvedev, *The Rise and Fall of T. D. Lysenko* (New York: Columbia University Press, 1969); David Jaravsky, *The Lysenko Affair* (Cambridge, MA: Harvard University Press, 1970); Valerii Soifer, *Lysenko and the Tragedy of Soviet Science* (New Brunswick, NJ: Rutgers University Press, 1994); Nikolai Krementsov, *Stalinist* Science (Princeton, NJ: Princeton University Press, 1997); and Nils Roll-Hansen, *The Lysenko Effect: The Politics of Science* (New York: Humanities Books, 2005).
57. Theodosius Dobzhansky to Ashley Montagu, 22 May 1944, Montagu Papers, American Philosophical Society, Philadelphia, PA.
58. Theodosius Dobzhansky to Ashley Motangu, 12 January 1961 (see n. 57). Dobzhansky was not alone in criticizing Montagu, and the Montagu Papers at the American Philosophical Society have responses from a number of zoologists and anthropologists who agreed with Dobzhansky. Montagu was not successful in popularizing *ethnic group* in anthropology, but, in a sense, he was on the winning side in that the use of the term *race* was increasingly seen as not having any biological meaning. See Nathaniel Gates, ed., *The Concept of "Race" in Natural and Social Science* (New York Garland, 1997).
59. See Gates, *Concept of "Race"* (see n. 58).
60. L. C. Dunn and Theodosius Dobzhansky, *Heredity, Race, and Society* (New York: New American Library, 1946).
61. *Origin and Evolution of Man*, 317 (see n. 42).

Race, IQ, and Politics in Twentieth-Century America

Hamilton Cravens

I.

In 1916 the New York attorney, racist, and conservationist Madison Grant published a book that was to be influential in the United States for the next two decades, *The Passing of the Great Race.*[1] The book went through many printings, eventually selling some 16,000 copies in this country alone. In that book, Grant famously argued that it was the Nordic race that was responsible for every outstanding accomplishment of civilization and for all the great civilizations. Race was the engine of history—and anthropology. There were three general races of mankind: the European, the African, and the Asian. Within the European race, there were three distinct races, each with its own merits or lack thereof. Thus the Nordic, according to Grant, was a race of soldiers, sailors, adventurers, explorers—nay, of leaders, rulers, organizers, and noblemen. They stood in total contrast to the essentially peasant stock of Alpines. The Mediterranean type or race had some artistic ability, and, when mixed with the Nordics, as in Egypt, Crete, Greece, Rome, and other such cultures, it produced notable civilizations.[2]

Grant's book resonated in a decade in which there was a tsumani of anti-immigrant and anti-black emotion in American culture and society, a high-water mark of popular racist feeling on the part of so-called white Anglo-Saxon Protestant Americans against those groups they thought inferior to them, especially immigrants from southern and eastern Europe and persons of color, especially African Americans, but truly any dark-skinned group. This was also the climax, the crescendo, of the progressive era's movement for the segregation of blacks from whites and of the attempts, largely successful in the former states of the Confederacy, to disfranchise black voters and to deny blacks and other persons of color anything beyond an inferior, second-class status in society. Indeed, so popular did Grant's work become that it was translated into numerous languages, including German, and it had a profound impact upon the Nazi Party's ideologues in the 1920s and 1930s. Indeed, Adolf Hitler himself wrote to Grant that his book was his bible.[3]

Grant was a prominent New York City attorney as well as an advocate of immigration restriction, eugenics, and conservation. A tall, handsome man and a natty dresser with a handlebar mustache, piercing eyes, and a straight

"Nordic" nose, Grant, as a champion of "scientific racism," fought with the Columbia University anthropologist Franz Boas over the problem of race. In 1918 Grant founded the Galton Society, named for Sir Francis Galton, the founder of the eugenics movement, a dinner and discussion club at which only those who agreed with Grant's racist views could attend or become members. Grant hobnobbed with important political figures, including Theodore Roosevelt and Herbert Hoover. He became deeply involved in the immigration restriction movement and was responsible for drafting some of the movement's laws successfully enacted by the U.S. Congress. Grant was indeed riding a tide of racist and restrictionist sentiment. His work and ideas had an impact on science as well.

Indeed, it cannot be emphasized too strongly that Grant was a symbol for an entire era, stretching from the 1870s to the 1920s and even into the 1930s, of open, brutal racism, segregation, and disparagement of persons of color as well as those from alien lands. If it was the case (and it was) that there were scientists who were opposed to these kinds of statements, not to mention the actions they condoned. One such scientist was the German-American professor of anthropology at Columbia University Franz Boas. Boas was relatively alone, not exactly a lone voice crying in the wilderness, but clearly someone without much support from the scientific or even the academic communities, even in the so-called "Progressive Era." Boas was indeed was part of a mere "corporal's guard" in comparison with the many loyal soldiers who marched alongside Grant and his ilk.

In the early 1920s there was a hue and cry—indeed, a hubbub—throughout the land over a new social technology and its implications for America's prospects as a civilization. This new social technology was the intelligence test. The specific event was the United States Army's use of intelligence tests during World War I to sort out men in its ranks into officer and enlisted men material, and then, hopefully, to fit men of specific skills (or lack thereof) into particular jobs in the service. In the national political arena, a hot and heavy debate over immigration restriction was brewing. In 1921 Congress enacted a temporary restriction law.[4]

This in turn provoked a new and more intense phase of the public debate. But now the champions of immigration restriction, who had claimed that the "new immigrants" from southern and eastern Europe constituted races below the American standard, could now point to the results of the army tests as proof—scientific proof, no less—of the truth of their case. In the spring of 1921, Professor William McDougall, a former British don and now Harvard professor of psychology, published a provocative book, *Is America Safe for Democracy?* Taking his cues from Grant, among other apostles of scientific racism, McDougall reviewed with dismay the results of the army tests, which had been referred to in the popular press as showing that the average intelligence age of American soldiers was not quite fourteen years. Clearly McDougall was not the most alarmist interpreter of the army tests, but just as clearly he believed that American society was on the brink of social catastrophe. America was developing with such speed that disaster was a genuine possibility because Americans were ignoring the fundamental laws of biology and psychology. Without proper scientific knowledge, Americans would never be the masters of their own fates. The army tests had proven the superiority of the "Nordic" races to all others, he intoned. He grimly concluded that if the flow of the inferior races from southern and eastern Europe were not stopped, the American experiment in democracy would fail.[5]

In February 1922 Mrs. Cornelia James Cannon created a public brouhaha when she claimed in *The Atlantic Monthly* that the army tests had shown the intellectual inferiority of immigrants from southern and eastern Europe. She exhorted the United States Congress to stop the flood of persons with such defective germ plasm into the country. The presence of so many genetically inferior voters endangered the very stability of the republic. After all, this blueblood matron of Harvard's Radcliffe Club of New York insisted, everyone knew that low intelligence caused people to vote and behave stupidly and without any sense of public responsibility.[6]

Another self-proclaimed blueblood, the Boston popular writer and eugenist Lothrop Stoddard, published a dire, pessimistic interpretation of the army tests later that same year. In *The Revolt Against Civilization*, he insisted that the army tests proved that there were races of "undermen"—a term that the National Socialists took up in their own propaganda in Germany—that would drag down Anglo-Saxon civilization. He was indeed a protégé of Madison Grant. "Civilization," he insisted, is " the result of the creative urge of superior germ plasm. Civilization is thus fundamentally conditioned by race." The inherited traits of a society's elite determined that society's history. Science had shown conclusively that intelligence was inherited, varied tremendously from one race to the next, and not modified by the social and cultural environment. Restriction of immigration was the only thing that would save America from certain disaster.[7]

In the fall of 1922, George B. Cutten made sensational newspaper headlines when, in his inaugural address as president of Colgate University, he insisted that one could not "conceive of any worse form of chaos than a real democracy in a population of an average intelligence of a little over thirteen years."[8] Cutten referred, of course, to the popularized findings of the United States Army's mental classification tests of its more than 2 million volunteers and conscripts during World War I. No sooner had the United States entered that conflict than Robert M. Yerkes, the brash young president of the American Psychological Association and junior member of the Harvard University faculty, offered the services of his professional colleagues to the army to help sort and select the vast numbers of new men coming into the army for the war. With a committee of professional colleagues, Yerkes devised two group tests: Alpha for literate men and Beta for illiterates (at least in English). Once standardized, the Army Sanitary Corps' psychological division, which Yerkes commanded as a major, administered the tests to large groups of men at one time at induction centers and military bases throughout the country.

After the war, Yerkes and his colleagues offered detailed analyses of the enormous mountain of data they had amassed. Perhaps the most elaborate scheme was that which Yerkes's friend Princeton psychologist Carl C. Brigham constructed in his 1923 book, *A Study of American Intelligence*. Brigham had worked with Yerkes in the army, administering the tests to soldiers. His book was a vigorous defense of the tests and of the notion of "Nordic" intellectual racial superiority over all other races. Brigham defended the army tests as reasonable, judicious measures of native intelligence. Of course, they were information tests. The possession of information, he insisted, was an indication of intelligence, and the tests showed that the average scores of various ethnic and nativity groups corresponded to the racial types that physical anthropologists had identified among white Europeans—Nordic, Alpine, and Mediterranean. Brigham knew there was no scientifically rigorous way to estimate the distribution of each type among immigrant groups. With great gusto, he invented his own estimates by calculating the percentages of Nordic, Alpine, and Mediterranean "blood" in each ethnic or nationality group. Then he attributed the higher average test scores of the Nordics to race and original nature. "The results which we obtain by interpreting the army data by means of the race hypothesis," Brigham wrote, "support Madison Grant's thesis of the superiority of the Nordic type, over the Alpine and Mediterranean."[9]

It should be borne in mind that the racist interpretation of the tests found support from another quarter: in 1904 a British psychologist, Charles Spearman, published a paper that was to be seminal in psychology, on what he termed the two-factor theory of the structure of intelligence. Just as those who devised the statistics of regression by reference to two elements—one determinate, the other indeterminate—so Spearman elected to interpret correlation between two factors as meaning there existed a common factor and in each variable a specific factor. Spearman decided that the dominance of positive correlations was caused by the presence of a general ability common to all kinds of performance, which he called *g*. Over time *g* came to be regarded by most psychologists as intelligence. The idea that there was a factor that was intelligence sustained the notion that intelligence was inherited as if it were a Mendelian unit character, or, in somewhat later terminology (e.g., the 1920s) a gene. Thus hereditarian psychology and hereditarian genetics went hand in hand so very conveniently.[10] But this was not the end of the matter.

As Cutten's address made clear, the army test results kicked up quite a stir. Three general findings in particular stood out like the proverbial sore thumb. They raised nasty suggestions about human inequality and racial differences. First was the finding that the average intelligence of the examinees, when translated into mental age scores, was slightly more than thirteen years. The average adult was supposedly sixteen years in mental age, and the recruits were at the upper level of being morons, or high-class mental defectives (in that day's language). Yerkes and his colleagues took it for granted that the tests measured native intelligence. The tests, they assumed, were perfectly reasonable and accurate instruments. Yerkes was a eugenist, as were most scientists who agreed with him. To them it appeared that their most dire predictions about the deterioration of the nation's polyglot racial population, especially its hereditary equipment, were scientifically proven.

Second, and flowing from the first controversial finding, the average mental ages of different nativity or ethnic groups varied, with those from foreign lands, especially southern and eastern Europe, and those with duskier skins doing much more poorly than the white, Anglo-Saxon Protestant examinees born in the United States or of northern and western European parentage. As Brigham interpreted the data in *A Study of American Intelligence*, the native-born white Americans and the Americans of northern and western European extraction or birth, or both, did better than those from southern and eastern Europe because they came from races that were superior; it was their superior "Nordic" blood that trumped the lesser "Alpine" and "Mediterranean" blood of those from southern and eastern Europe. As Yerkes, Brigham, and their allies put it, there was a genetic basis that made certain races mentally superior to others. And, of course, persons of color—especially African Americans—sank below even the lowest-scoring Italians and Poles.[11]

Last, but hardly least, there was a very low average score for American blacks, even lower than for any white immigrant group, with the average IQ for African Americans in the mid-80s, several points below that of Italian and Polish immigrants, and, of course, one full standard deviation, or fifteen points, from the norm of 100. Again, Yerkes, Brigham, and their allies chose to view these average scores as the result of inferior racial endowments of black persons as compared with whites of any nativity group.[12] And the army tests had more than an intellectual influence. They added ammunition for those in public affairs and politics who were agitating for immigration restriction, although most historians agree that immigration restriction laws would have passed in the United States Congress regardless of whether there were the army tests or not; in fact, Congress had enacted a temporary law in 1921, just a whisker before the army tests won widespread notice, and the National Origins Act, which Congress passed in 1924, set out a permanent quota system of about a quarter million per year based on ascribed national origins. Those who could claim northern and western European countries, such as England and Germany, as their ancestral homelands had many more places awarded than those from southern and eastern European lands, such as Italy, the Soviet Union, or Greece or other Balkan nations. The laws were made permanent in 1929, and they were changed to eliminate so-called racial or nativity quotas only in the mid-1960s, at the behest of President Lyndon B. Johnson.

Yet long before the immigration laws were revised to eliminate the racial quota system, the intellectual superstructure of the racist ideas emanating from the army tests collapsed, as Brigham conceded in 1930.[13] Brigham was under heavy attack in various reviews and critiques by other scientists, including the distinguished psychologists Edward G. Boring of Harvard and Frank N. Freeman of the University of Chicago, and noted anthropologists such as Boas and the American Museum of Natural History's Margaret Mead.[14] The journalist and public philosopher Walter Lippmann also penned a long series of criticisms of the army tests and their racial interpretation in *The New Republic*, which created quite a stir throughout academe and up and down the eastern seaboard.[15] The criticisms of the Nordic view were multiple, but all the critics focused on the culture-bound character of the tests. They also insisted that a number of the findings could be used to support an environmental and liberal interpretation of the tests and the examinees' intelligence. For example, northern African Americans scored higher average IQs than blacks from the south, which Yerkes, Brigham, and others tried to explain with a selectionist argument. They insisted that the more intelligent blacks, which by definition they assumed had a higher admixture of white ancestry, sought better opportunities in the Northern states. Yet there were difficulties with this theory, not least of these was that African Americans residing in Northern states had significantly higher average scores than many southern whites.

Other evidence existed, in plentitude, that the racial interpretation of the army tests was dubious. Thus with regard to the different average scores of immigrant groups, those groups who had been in the United States the longest, regardless of their ancestral homelands, did better than those who had been here the shortest length of time. Those residing in the United States for twenty or more years had an average of almost 13.75 years in mental age and scored higher than examinees in general, whereas those who had been in America for ten years or fewer scored an average of 11.70 years in mental age, and those who had been here five years or fewer managed an average mental age of only 11.29 years. But more than length of residence mattered. The measured quality of a state's schools was crucial as well, and these two factors acted together to boost the longer-residing immigrant groups above the average mental age score for examinees in general and for certain "Nordic" groups (for example, white Southerners) as well. Furthermore, as the tests and other tests were discussed by a more diverse group of scientists, including some of immigrant and Jewish background, it turned out that members of certain groups, including immigrants from Japan and China and eastern European Jews, scored significantly higher on the Stanford-Binet tests than did the average white, Anglo-Saxon California middle-class schoolchildren upon whom the Stanford-Binet tests had been standardized in the first place.[16]

There were other considerations as well that ate away at the legitimacy of the hypothesis of innate racial superiority that Cutten, Yerkes, Brigham, and many others had invoked. For one thing, the genetics community in the 1920s was already backing away from supporting extreme hereditarian interpretations of inheritance, development, and variation, and, above all, distancing itself from the extremists among the eugenics movement. In the 1930s and 1940s geneticists in particular and evolutionary biologists in general were proposing wholesale revisions of the notion of superior and inferior races. Their criticisms and theoretical stances were complex, but they may be summarized as follows: Inheritance was reduced from the all-powerful agent to an important element that interacted with environment in the development of individuals and species. Evolutionary theory admitted contingency, randomness, drift, and unexpected consequences. And evolutionary theorists such as Theodosius Dobzhansky of Columbia University insisted that culture worked with nature in evolution and that there was cultural as well as biological evolution. Most of these scientific statements were made before the Second World War and the political reaction against National Socialist racism and genocide.

By the early 1920s Franz Boas and his school of anthropologists were gathering together a powerful critique of racial mental testing and the entire notion of a hierarchy of superior and inferior races. From the mid-1920s until after World War II, they became very strong critics of the racial interpretation of mental tests. Since the 1890s, when he was at Clark University, Boas had been taking anthropometric measurements of various groups of Americans, including members of particular racial and ethnic groups. In time Boas became one of the most sophisticated physical anthropologists in North America, if not also in Europe. His advanced liberal political views, especially his arguments that all races were culturally and intellectually equipotential, did not sit well with a number of eminent American scientists or even American anthropologists. When, in 1919, he publicly accused the Wilson administration of using government anthropologists as spies in Mexico, as historian George W. Stocking, Jr., has ably related, his fellow anthropologists in the American Anthropological Association censured him for a lack of patriotism. Within a few years, these tensions subsided, in no small measure because the demographics of the anthropological profession changed from being dominated by government scientists, who were often very conservative men who agreed with racists such as Grant and Stoddard and committed to materialistic schemes of evolutionary ethnology based on notions of a hierarchy of superior and inferior races, to academic scientists, mostly whom Boas had trained and who were in any event professors of left-liberal political views, and, in numerous instances, of Jewish ancestry. The discipline's center had moved from government museums to research universities, with profound implications for new points of view, as Stocking has elegantly underscored.[17]

In numerous rigorous monographic studies, Boas outlined the case against a theory of superior and inferior races from the perspective of physical anthropology. In 1903 Boas contradicted the work of English biometricians Galton and Karl Pearson with regard to the shape of the head form, which more nearly followed the Mendelian expectation and not some presumed midparental type, as these founders of the science of biometrics assumed.[18] Six years later Boas insisted that the biological and anthropological issues that had been brought to the fore by the mixture of various European nationalities with Indians, blacks, and Asians had not been rigorously or carefully defined or studied. After carefully reviewing the history of the mixing of European strains in the Old World and in the New World, he insisted that "we may dismiss the assumption of the existence of a pure type in any part of Europe, and of a process of mongrelization in America different from anything that has taken place for thousands of years in Europe." In other words, "the concern" many felt that "racial purity" be continued in America "is to a great extent imaginary." Indeed, physical types changed in response to the environment. Here Boas was referring to statistics he had been collecting for the United States Immigration Commission, which showed definitely that the head forms of the children of eastern European immigrants were more like those who had long been in America and less like those of their grandparents.[19] While it might be true that blacks were a different race than Europeans, there was no evidence that this meant they were inferior to whites. Difference, in short, did not add up to inferiority or superiority.[20] In reviewing the work of Professor R. Livi of the variation of cephalic indexes of Northern and Southern Italians, who represented different head forms, again the conclusion was the influence of the environment. If a stable physical trait such as head form could alter with a change in environment, Boas insisted, surely more malleable intellectual and social characteristics could also change with a new environment.[21]

But there was more ammunition in the Boasian armamentarium. It was one thing to turn upside down the racist thesis that as the human body had evolved and was inherited, the same was true of the mind. It was another to take the argument of mid-nineteenth-century anthropologists who, following Herbert Spencer and others, insisted that all mankind followed definite evolutionary stages from barbarism to savagery to civilization. Here Boas and many other anthropologists, including many of his doctoral graduates such as Alfred L. Kroeber, Robert H. Lowie, and Alexander A. Goldenweiser, claimed that what mattered was that each people had its own historical experiences bounded by various cultural relations, symbols, rituals, and other communal shared patterns of behavior. As Lowie put it, all cultural behavior and phenomena came from prior cultural behavior and phenomena; biology, psychology, physics, and other determinants of the natural sciences were irrelevant to the cultural history of any given people. Accordingly, the entire edifice of a hierarchy of superior and inferior races collapsed, a last shred of Aristotlean thinking, according to which the data of the universe have differing moral and cultural meaning. The mentality of the modernist philosophy of science, according to which the data of nature are just data, without moral or cultural meaning, had finally penetrated anthropology.[22]

By the early to mid-1920s, psychologists were picking up on the difficulties, methodological and empirical, of the racial interpretation of mental tests. This was especially the case with psychologists who were social workers or in other cases with no traditional academic professors of psychology. In January 1924, Maurice B. Hexter of the Federated Jewish Charities of Boston and Abraham Myerson, a professor of neurology at Tufts University Medical College, published a slashing critique of Brigham's work in *Mental Hygiene*, a scientific journal for psychiatrists and social workers. They accused Brigham of multiple errors. Brigham had practically insisted that the army tests accurately measured innate intelligence, thus ignoring their culture-bound character. And the mental ages of the army recruits had been calculated on the Stanford-Binet test, which itself had been standardized on a statically inadequate number of cases. They also noted that Brigham's own data showed an increase in mental age with longer periods of residence in the United States and insisted that Brigham's racial explanation was wrongheaded. That fact simply showed that the longer any person was exposed to the American environment, including its schools, the better the person did on the tests. Brigham's work had other careless statistical blunders as well; for one, there were too few foreign-born individuals in the army sample to be valid in most instances, and therefore the majority of mental ages thus calculated for immigrants were seriously out of proper adjustment. All the army tests showed, they noted, was that recent immigrants were slower in their performance on the tests and had a lower score than either the native-born or long-resident immigrants; the rest was unjustified assumption. The army tests, they concluded pointedly, "merely measures the inherent disaster attendant upon the transplantation of an adult group from one environment to another."[23]

In the mid-1920s, in a critique of racial mental testing, Mead, then one of Boas's advanced doctoral students, insisted that a sound methodology had not yet been created for such assertions. She reviewed the work of mental testers who dealt with three special issues concerning racial intelligence. What of the problem of racial admixture and intelligence scores? she asked. The laws of inheritance for intelligence did not exist. Hence the only empirical method would be to match, if possible, the school records of children of various races. Yet even here there were serious problems, for it would be difficult in the extreme to sort out the variations in social environments of particular groups. Nor could one, as some had tried, use an estimate of black-white mixture by a visual inspection of skin color. Skin color was not a viable measure of racial mixture.

And what of attempts to measure the influence of social status on test scores? The best work, which Ada Hart Arlitt of the University of Cincinnati did, involved two studies: one a comparison of black and Italian school children as compared with white native-born pupils; the other in which she deployed a crude occupational rating scale for the families of the three groups. In both instances, when children of similar social rankings were compared, the IQ scores were within a few points of one another and not more than twenty points apart. And, finally, language handicaps, especially for children whose native language was not English, were a striking issue and an authentic and disabling factor. Mead concluded that "extreme caution" must be used in any attempt to make conclusions on the relative intelligence of different racial or nationality groups on the basis of tests, "unless a careful consideration is given the factors of language, education, and social status," and further adjustment must be made for the influence of different cultural habits and attitudes.[24]

Other anthropologists joined the fray. Alfred M. Tozzer of Harvard, for example, accused the testers who claimed that there were innate racial differences in intelligence of confusing culture and nature, and pointed out that the army tests did not show clear-cut racial differences (i.e., between blacks and Italians, who were too close together in average mental age) and, in any event, Northern blacks had far surpassed Southerners, both white *and* black. Melville J. Herskovits, a physical anthropologist at Northwestern University, accused Brigham of using unscientific racial and social categories. And so it went. According to Boas and other anthropologists, the tests were culture-bound, inaccurate, based on unscientific categories and assumptions, and so contaminated by sloppy methods as to be without merit.[25]

In the late 1920s psychologists began to turn en masse against the racist version of racial mental testing. It began as a trickle of critical statements in the early 1920s, with, for example, that of Joseph Peterson of the George Peabody College of Education, who had served with Yerkes in the army, testing soldiers' intelligence or lack thereof, and who had even belonged to a eugenics society for a few years. Peterson nevertheless insisted that the racial testers were using shaky methods and making unwarranted assumptions. The tests, he claimed, were at least partly culture-bound.

Another complication for the racist interpretation of mental testing came again from the psychometricians, who focused again on the structure of intelligence, as Spearman had done with *g* as intelligence in 1904. This was factor analysis, developed by Godfrey H. Thompson at the University of Edinburgh, Cyril Burt at the University of London, and L. L. Thurstone at the University of Chicago. Factor analysis assumed that there were a multiplicity of factors to intelligence and that they were intercorrelated and could be computed into as many separable factors as could be calculated. From this perspective, especially insofar as Thurstone was concerned, there was no such thing as a unitary element of intelligence; there were many intelligences. That made the genetic theory of racial intelligence more difficult to sustain, to say the least.[26]

David A. Wechsler, who came to the United States from Romania at the age of six with his parents, also questioned the racial interpretation of the army tests. A brilliant young man who raced through the City College of New York, graduating in 1916 at the age of twenty, Wechsler did his graduate work at Columbia University, the University of London, and the University of Paris and finished his PhD at Columbia in two years. He published several seminal articles on mental testing and questioned the racial interpretation of the tests, insisting that cultural and educational handicaps most likely explained why recent immigrants did not do as well as native-born Americans on the tests. Wechsler went on to become a major figure in the post-1930s mental testing field and a strong advocate of environmental and cultural interpretations of mental testing. After the Second World War, he had much to say about the structure of intelligence. But an important turnabout of psychologists on racial mental testing took place from 1925 to 1930. In 1930 two important pioneers of the racial interpretation of the tests, Brigham, and Thomas R. Garth, renounced their former views as unscientific and unsupportable.[27] These were devastating defections from the hereditarian camp.

Clearly the larger political context was changing somewhat in the 1920s, especially in the later years of that decade. Immigration restriction by "racial" quotas became federal policy and law, thus giving the immigration restrictionists and racists a huge victory. The Ku Klux Klan began to lose ground in a number of states as the corruption of some of its leaders gained national attention, but the campaign of white Southerners to clamp down on African Americans through a wholesale regime of segregation emerged victorious by the decade's midpoint. It may appear ironic that these political victories of racist conservatism cooled off the tensions of the day, but that is what apparently happened. In such a different environment, those who criticized racist doctrines could speak with more freedom, for the racists and the segregationists had "won" their political struggles. Yet it was not until the 1950s and 1960s that segregation and racism went on the defensive in American culture and politics. Until then, segregation and institutionalized racism were thoroughly ingrained into many aspects of American life, including education, employment, housing, military service, and voting.[28]

II.

How did the ideas of race change in science? The identity of the group to which an individual "belonged" was still the key to that person's essential character and prospects. That much had not changed, at least not in any major way. What had changed—and rather dramatically at that—was the manner in which scientists explained the reasons for, or the causes of, group identity. Since the 1870s it had been all but universal to explain why races differed from one another by invoking biological mechanisms, processes, and structures. There was always an appeal to the power of biology, of innate characteristics passed along through the processes of physical inheritance, even though scientific knowledge before 1900 about how physical inheritance worked had not proceeded much beyond the cliché "like begets like." The rapid development of experimental zoology and botany and the development of the chromosomal theory of inheritance, when coupled with the "rediscovery" of Mendel's laws of assortment and recombination of unit characters in the two or three decades after 1890, led to a confident scheme of superior and inferior "races" based on the most up-to-date science. That the psychologists reinforced the biologists with their technologies of mental measurement on the question of innate racial differences simply sweetened the pot.

What crystallized in the interwar years was what came to be dubbed the "dual inheritance" theory—that is, the notion that man was the product of biological *and* cultural or social inheritance and therefore man was the only species on the planet that had undergone both *biological* evolution, meaning his physical and biological characteristics and behavior patterns, and *cultural* evolution, meaning the development of language, abstract thought, inventions, rituals, literature, art, and the like—and the stunning variety of human cultures that existed or had existed in human history.[29] It was the social disciplines—clearly anthropology but no less certainly sociology—and the other social sciences whose practitioners had contributed mightily to this intellectual shift. There were roots of this change in politics as well, as in the liberal and Marxian critiques of scientific and popular racism. There were various expressions and formulations of these perspectives, popular and scientific. There was some question as to how far to take the cultural perspective, at least in the 1930s. But all agreed that nature and culture interacted, as if oil and water could mix. Both mattered; neither could be dismissed. The prevailing view among leading geneticists and anthropologists, especially from the 1940s on, was that although man was as subject to the laws and processes of biology in his physical aspects, as was any species, what made him unique was his ability to create culture in all its manifestations.[30]

Put simply, in terms of scientific ideas, the change was from a formula of nature *over* nurture to nature *in concert with* nurture—that is, from *domination* to *interaction*. In terms of debates and discussions about science and race, a perfect example of the change since the army tests was over the question of "selective migration" of Southern African Americans to the North, especially during and after World War I. One of the results of the army tests, which Yerkes, Brigham, and others noted, was that Northern African Americans achieved higher average scores than did African Americans from the South. Those who wished to emphasize race as an explanation, such as Yerkes and Brigham, insisted that it was the "better" African Americans who migrated to the North because they had the superior intelligence and character to do so. Others believed that it was the superior cultural and social environment of the urban North, especially its superior schools, that explained the difference in average scores.

Here the contribution of Otto Klineberg, one of Boas's PhD students, was crucial. In a compact monograph titled *Negro Intelligence and Selective Migration*, published in 1935, he systematically examined each aspect of problem. First Klineberg reviewed eleven studies of the average IQs of black children in the North and thirteen comparable studies of Southern African American children that had been done since the army tests. The average IQ for all the Northern groups was 86.3. For the Southern groups the average was less—79.6. Given that a score of 100 was supposed to be average or normal, seemingly there was a problem with African American intelligence, at least as measured by IQ tests. Yet it was clear that the kind of test used made a difference. The group tests yielded a comparison of 82.9 among the Northern children and 76.8, on average, for the Southern children. With the Stanford-Binet tests, there was almost no difference—88.2 and 88.5, respectively. The Stanford-Binet test was an individual test that took almost three hours; the group tests were given to substantial numbers of children, comparatively speaking, and took perhaps forty-five to sixty minutes. After reviewing a large body of evidence, Klineberg concluded that the Northern blacks did better on average than Southern blacks, and that the Northern blacks approximated the scores of the Northern whites. "There is, in fact, no evidence whatever in favor of selective migration," he insisted. The school records of those who moved north did not demonstrate any superiority over those who did not. The intelligence tests demonstrated no superiority of those who had just moved to the North over those of the same sex and age who remained in Southern cities. And there was definite evidence that "an improved environment, whether it be the southern city as contrasted with the neighboring rural districts, or the northern city as contrasted with the South as a whole, raises the test scores considerably," and that this rise in intelligence was roughly proportionate to the length of residence in the more favorable environment.[31] And Klineberg sounded an increasingly common theme of 1930s social science about racial minorities: that the American environment *damaged* minority children, made them believe themselves to be inferior, which in turn became a self-fulfilling prophecy. This argument first emerged clearly at the 1930 White House Conference on Children, and it was soon a leitmotif in works such as John Dollard's *Caste and Class in a Southern Town*, published in 1937, and other such works that followed, especially Gunnar Myrdal's *An American Dilemma*, published in 1944.[32]

By the 1930s and 1940s the racial interpretation of IQ testing was clearly on the defensive, at least in the scientific community and, quite probably, in certain circles of educated whites, especially but not entirely outside the South. There were powerful scientific reasons for this. Aside from those we have seen, the scientific case for a hierarchy of innately superior and inferior races had been clumsily and ineffectually made from the 1900s to the 1920s. Several powerful political tides sustained the scientific argument and sheltered it from out-and-out attacks before the 1920s. One was the political campaign to enforce racial segregation in the post-Reconstruction South, which, as Howard Rabinowitz has reminded us, was the consequence of emancipation. This Southern Negrophobic movement entered its final stage with the disfranchisement campaigns there, starting (mainly) with William McKinley's victory and William Jennings Bryan's defeat in 1896 and coursing through Southern politics for the next two decades.[33] The so-called Free States, where slavery had been eliminated after the Revolution, adopted segregation for African Americans before the Civil War as the only appropriate public role for the former slaves. The second, which paralleled and reinforced the first in many respects, was the immigration restriction movement, which operated largely outside the South and in those parts of the nation where immigration restriction was a burning issue—notably the corridors of states alongside the Atlantic and Pacific oceans and in any place, such as Colorado or Iowa, where immigrants existed in large enough numbers to annoy and alarm the native-born whites and therefore experienced stigmatization. A third movement, not involving the same numbers of persons but nevertheless of approximately equal intensity and ferocity, were the efforts to segregate Native Americans as well as persons of Asian and Hispanic descent. Indeed, there were also strenuous efforts to segregate European immigrants from middle-class white society. Together, and in concert with the Progressive movement, many of whose leaders and members championed segregation of the minorities, these political tidal waves made the case for scientific racism appear stronger than it was.[34]

In the interwar years, the scientific theory of innate racial differences came under attack within the scientific community and, to a lesser extent, from some white liberals, especially Jews and some Quakers, Episcopalians, Unitarians, and members of other high-status "liberal" or "progressive" churches. Then World War II, especially the revelations of the Nazi slaughter of the Jews and other "racial" groups—Russians, Poles, Gypsies, and others—helped bring ideas of superior and inferior races into moral disrepute some years *after* World War II, especially as the story of the Holocaust unfolded in American life.[35]

In the interwar years, America was still racially segregated. There was still belief and action throughout the land that served as powerful evidence that most white Americans believed that members of minority groups—persons of color, that is—were simply inferior to them in every way. President Harry S. Truman supported civil rights in his 1948 bid for the presidency, thus splintering his own party into intransigent Dixiecrats and regular Democrats, and in 1949 he ordered the desegregation of the armed forces. Yet these steps, as courageous as they were, barely made a dent in segregation's formidable armor, especially when we consider its expanse in the society at large. Indeed, segregation was the law of the land, underwritten and guaranteed by the federal constitution.

And, indeed, within the world of the human sciences, most scientists believed that if they could not openly support the notion that science "proved" some races were superior to others, they could insist that heredity prevailed over environment in the making of human nature, conduct, and intelligence. What was the evidence that psychological testers insisted was crucial that the IQ was innate and constant in each and every individual person? They cited six reasons:

1. From the beginning of the Stanford-Binet-type tests, performance generally improved with age, thus suggesting that an authentic developmental process was being measured. This quickly led to the adoption of the intelligence quotient (IQ)—that is, the ratio of mental to chronological age.

2. The average IQ from age to age was highly constant. This resulted, of course, from the test's design, in which one-fourth of the children tested were assumed to fail because they were stupid, one-half were in the middle range of test scores, and one-fourth were in the upper range. That the mean IQs for groups was highly

constant proved only that enough consistency existed to make average performance approximately replicable from one examination to the next. That said absolutely nothing about the constancy of the IQ in individual children.

3. The IQs of individual children tended to be consistent on retesting. When there were exceptions, the testers explained them away as measurement errors.

4. Scores on these tests tended to correlate with one another rather closely, leading psychologists to believe that there was a particular factor for intelligence, which they dubbed g; this was the mental horsepower an individual displayed, and when multiple factor theories of intelligence came along, they substantiated the belief in consistency in test scores, which, in turn, "proved" that g existed. There was a fair degree of circular argumentation here.

5. Intelligence test scores tended to be fairly successful in predicting later success in school, as well as in one's line of work, especially for those of a technical or professional character in which education and training of a high level was a minimum entrance requirement.

6. There was evidence of hereditary determination of intelligence, usually through studies of the intellectual likeness, as measured by the coefficient of correlation, of related individuals reared in the same or different environments. The evidence seemed the most convincing in research in which identical twins were reared apart, if it could be shown that a dissimilar environment did not produce dissimilar intellects in persons with presumably identical genetic endowments. But such investigations were also tried for fraternal twins and other close but not identical relatives, and the coefficients matched for each type fairly closely.

Another source of evidence came from studies of human development, according to which there was lawful maturation of inner mute anatomical structures that were themselves the result of genetic inheritance. The so-called maturation theory constituted a mainstream doctrine among virtually all developmentally oriented scientists. It also supported the notion of the fixed, innate IQ.

An important illustration that nature versus nurture was still a problem in the interwar years came with the work of the Iowa Child Welfare Research Station on IQ. Founded in 1917 as a result of Progressive reform pressure, the Iowa Station's scientists took as their assignment the study of young children's physical and mental growth. In the station's first twenty years, the scientists assembled a veritable mountain of statistics of physical growth. By serendipity they developed an important and eventually controversial program of mental measurement, starting with the need to accurately measure the IQs of children to be adopted from state orphanages to ensure that prospective parents adopted children of normal intelligence. Shocked by the rather low average IQ of many orphanage children, the station's director established an all-day preschool at the state orphanage, located in Davenport, some fifty miles east of the station. Children in the preschool were compared with those who remained in the orphanage without the benefit of preschool. The several studies that sprang from this work, and that of other groups at one of the station's preschools in Iowa City, emphasized that the average IQ of the preschool children gained perhaps a few points over a year or more, but the average IQ of those not enrolled in the Davenport preschool declined, and precipitously so, from the dull normal range to borderline retarded or even worse. At the university's preschool, the average IQ rose after about a year and then rose above the expected variation upon retesting.[36]

In 1939 Marie Skodak published the startling results of her graduate thesis. She followed the mental development of 154 children from the Davenport orphanage who were adopted into foster homes under the age of six months, and another 54 from the same orphanage under the age of two and one-half years who were also placed with families. All were initially tested in the orphanage and scored in the so-called feeble-minded or retarded range, with IQs ranging from 50 to 80 IQ points. Yet after a year in the foster homes, all children tested in the 110–130 range—above-normal IQs that were congruent with their foster parents' and well above the discouraging range the very same children had when in the orphanage.[37] This was a spectacular rise and one that went against seemingly settled conventional wisdom among psychologists and educators on the innateness and fixity of the IQ.

But there was more, and it was even more shocking to critics of the Iowa Station work. And it was the result of serendipity, of happenstance, indeed. Harold M. Skeels, the young psychologist who had worked with the Davenport preschool and was a specialist in the mentality of so-called feeble-minded persons, found more ammunition for his argument that the orphanage was literally degrading the children's intelligence the longer they stayed there—that, in short, the orphanage was manufacturing morons. As a matter of routine, two retarded infants were transferred from the Davenport orphanage to the state institution for the feeble-minded in Glenwood, Iowa;

their IQs then were respectively 46 and 35 Binet points. The Glenwood facility was overcrowded, so the two infants were placed in a ward with female inmates ranging in chronological age from eighteen to fifty years and in mental age from five to nine years. Six months later, on a routine visit to Glenwood, Skeels was astonished to find that the two infants had much-improved behavior, suggesting a different IQ level; upon retesting, this was indeed the case. The two infants, now aged eighteen and twenty-two months, tested at 77 and 87 Binet points, respectively, a spectacular rise of 31 and 52 points. Twenty months later, Skeels tested them again; the younger child tested at 95 points, the older at 92, both in the normal or average range. Skeels ordered their transfer back to the Davenport orphanage, where they were placed for adoption.

Skeels explained this phenomenal development as the result of the intensive and special attention they received from the women in the ward. Put crudely, the "moron maidens" helped the two infants achieve a normal level of mental development. Skeels followed up this realization with a more systematic experiment. He arranged for the transfer of another thirteen feeble-minded infants and toddlers from the Davenport orphanage to the Glenwood home, to be cared for under the same regimen as the two infants. These children were very severely retarded—too much so for adoption purposes. Their mean chronological age was twenty months, and their mean IQ was about 65 points, with a range of 35 to 89. Over the next eighteen months, they were compared with a control group of children at the Davenport orphanage who were the same age and borderline normal. Again, the results were startling: Those in the experimental group in Glenwood made an average gain of about 28 IQ points, and those in the control group in Davenport suffered a mean loss of 26 IQ points.[38]

The ensuing controversy about these and other studies of the "wandering IQ," as it was sometimes derisively called, was relatively short-lived. By the early 1940s those champions of the mainstream view that one's IQ was innate and fixed at birth had carried the day. It might be instructive to understand the kind of evidence the attackers of the Iowa Station scientists' studies had, or at least thought they possessed. In those times, mental testers reported group averages. They did not, as a rule, track individuals. They also tended to report scores of children above the age of six and below that of eighteen. This was the conventional wisdom, and it amounted to a group portrait, frozen in time; the statistics were all averages. So long as a tester followed this procedure, the results were, as it has been pointed out earlier, roughly the same upon retesting. There was, of course, no evidence from genetics or biology that there were genes for intelligence located on this chromosome or that. Much was assumed, perhaps too much, as it turned out.

What the Iowa Station scientists had done, without understanding what they had done, was test and report on the IQs of *individual* children over time (often a year or more). The Iowa Station scientists tended to confuse group and individual performances, but the spectacular rises and falls in IQs they reported in their studies were all of individual children. There were some problems with their procedures. For example, they used the so-called matched pair method, in which each individual child in the control and experimental groups was evaluated on the basis of a relatively small number of traits, and when a child left one of the groups, as had to happen with threadbare little ones waiting to be adopted, the scientists looked for replacements who "matched" those who left in those few traits. Any child might be said to have many more than a half dozen or so traits that could be relevant. Yet even with this rather serious methodological problem, what Skodak, Skeels, and their colleagues found was so electrifying and so seemingly contrary to the conventional wisdom that a highly emotional controversy ensued, and, not surprisingly, the Iowa Station scientists lost the battle—and the war.[39]

Hence the possibility that the Iowa Station work or other studies of preschool children as individuals might have challenged the strong hereditarianism in the behavioral and social sciences, and therefore possibly might have undermined faith in innate racial differences, was never realized in the interwar years. There were other studies at the time that suggested that the IQs of very young children could be dramatically depressed or boosted; but in each case the scientists involved were specialists in early childhood research and education, and they were often female scientists who were research associates, without regular faculty status and not located in traditional psychology departments. These factors implied to many psychologists that these findings were suspect or even unscientific.

Yet matters did not rest there. The Second World War brought with it new developments that altered, to some extent, the existing cultural, political, and social aspects of race relations and majority attitudes toward racial minorities. By 1960 or so, the United States was still a deeply segregated society with daunting, widespread racism literally in the bones of its major institutions; nevertheless, important cracks and fissures had appeared in these structures. After all, racial stigmatization and segregation were far older in America's past than the scientific racist theories that emerged after the mid-nineteenth century. Racism had a political character long before it had a scientific rationale or justification. Hence changes in the politics of race mattered a good deal, and politics and the law were, of course, intimately related in America. During the war, with full employment available to them, African Americans found jobs in defense factories and other civilian pursuits and served in the armed forces. In some instances—the Congress of Industrial Unions and the automobile and steel industries, for example—were organized by the industry, which effectively integrated African Americans and whites. The armed forces remained segregated until President Truman ordered their desegregation in 1949, but African Americans in the war effort learned that theirs was a war against racism and tyranny that resonated when they returned to civilian life.

In 1944 the liberal Swedish economist Gunnar Myrdal published a massive study, *An American Dilemma: The Negro Problem and American Democracy*. The Carnegie Corporation initiated and underwrote the study, and selected Myrdal to run the project because it wanted an objective observer of American race relations. After six years of research and writing by a team of distinguished American scholars, the Swedish economist imposed a particular view of American racial relations. The racial problem, he insisted, was a problem with white racist attitudes and patterns of behavior about blacks. The "Negro problem," he insisted, was a white problem, a problem in the white mind. It was American whites who would have to solve the problem. Blacks were almost helpless to do much about the racial problem. The whites had all the power. And, he insisted, whites were divided about what to do about the problem. Myrdal insisted that the American creed of equality and fairness, which was inherited from the eighteenth-century Enlightenment, conflicted with American racial attitudes and with the color line. Resolution was essential for America, which would emerge from the war as an international leader; its skirts must be clean on the race issue.[40]

To a degree, Myrdal also invoked an argument from African American and white social scientists working in the new field of "culture and personality" such as John Dollard, Allison Davis, and E. Franklin Frazier, who insisted that American racism and segregation distorted black institutions such as the family, the community, and the school with damaging effects on the self-confidence and self-esteem of African Americans virtually from cradle to grave, thus limiting in cruel and artificial ways what they could accomplish in life—in school, in the neighborhood, on the job, on the athletic field, and even in religion and politics and in relations with their fellow human beings. This idea of the "vulnerable child" had been around at least since the 1930 White House Conference on Children, and had gathered force in the late 1930s and early 1940s. It was becoming a standard answer for why African Americans did not measure up to white standards in education, employment, crime, and many other social indicators.[41]

By the early 1950s, the National Association for the Advancement of Colored People was using this sort of social science testimony of cultural and psychological deprivation and oppression, rather than relying so much on socioeconomic comparative statistics of the circumstances of blacks and whites, in the various lawsuits it was pursuing to force the desegregation of public schools. The teams, led by attorney Thurgood Marshall, followed several different lines of argument: segregation was arbitrary, since science had proven that there were no innate racial differences in intelligence; segregation was disruptive to social relations because it blocked communication between the races; and segregation had damaging effects on the personalities of black children, which impaired their ability to profit from their education, and it had immoral and damaging effects on the personalities of white children. Desegregation also had its advantages, by improving race relations and being practical, as the integration of graduate and professional education had clearly demonstrated.[42] Kenneth B. Clark and his wife, Mamie P. Clark, Columbia-trained African American psychologists who became perhaps the most closely associated with the argument that segregation damaged black children, argued in 1950 that when African American children were asked to select what skin color they liked better, brown or white, they usually chose the latter. "It is clear that the Negro child, by the age of five is aware of the fact that to be colored in contemporary American society is a mark of inferior status," they concluded in the latest of their studies of the problem from several different angles.[43]

On May 17, 1954, the United States Supreme Court handed down its epochal decision on school desegregation in *Brown v. Board of Education*. What was remarkable about the decision was the Court's reliance on the social science theory discussed above—the so-called vulnerable child hypothesis. This was the first time the justices had ever used explicit social science theory in a case; in fact, in the famous footnote 11, Chief Justice Earl Warren referred explicitly to the Myrdal study and its allied references. The Court ordered desegregation in the public schools, sidestepping ruling unconstitutional the Court's approval of "separate but equal" in *Plessy v. Ferguson* in 1896, which, the justices insisted, only applied to the unimportant area of transportation; education mattered far more.

For the next several years, the character of American politics changed, and dramatically so. The most important domestic development was the growth of the civil rights movement. By the early 1960s the racial climate, like that of the larger culture, was changing dramatically, moving away from conformity to group and tradition and toward individualism, self-determination, and self-expression. This would have an important influence on the history of American racial mental testing.

III.

Perhaps not surprisingly, racial mental testing always went back to children—specifically to preschoolers. In 1961 University of Illinois child psychologist J. McVicker Hunt published an important attack on the mainstream dogmas of the field—the ideas of the innate fixed IQ and the maturation theory—in *Intelligence and Experience*, an overall assessment of the latest developments in the field. He expertly discussed the scientific evidence for and against each idea. He also discussed the work of the Swiss child psychologist Jean Piaget at some length. Hunt did not dismiss biological inheritance. He argued, however, that experience directly shaped and influenced the brain's development. Development was neither automatic nor genetically programmed. If organized structures existed as a part of the endowment of a species, it did not necessarily shape the development of an individual—or any number of individuals, for that matter. Experience was hardly linear. There were multiple—although clearly not infinite—possibilities in development.

Hunt wrote in a particular temporal context. American culture and society were moving away from the group conformity and determinism of the interwar years to our own age of individualism, fragmentation, and *n* possibilities and dimensions, as well as individual choice. These changes came early to popular culture, with the development of asymmetrical rock-and-roll music and the replacement of symmetrical popular music, and they spread to education and to public policy before the 1960s were over. One of the consequences of the liberalism of the Kennedy-Johnson years was the federal Head Start program, whose champions sought to take disadvantaged children—often African American children—and provide them with a special compensatory preschool education to boost their academic and social performance to something approximating a middle-class level so that they could benefit from elementary school on par with other children. The Iowa Station's work now had its heyday in national public policy, whether anyone really knew it or not. And cognitive psychology, emphasizing the cognitive development of the individual, became all the rage among psychologists. Hence compensatory education made sense in terms of science as well as public policy.

In this context, Harold M. Skeels joined the fray. After being discharged from the armed services following World War II, Skeels joined the federal government as a clinical psychologist. He decided in the later 1950s to do a follow-up to the original study of the "moron nursemaids" he had done with Harold Dye just before the war. On his own time, Skeels tracked down all thirteen in the experimental group who had had the special attention of the Glenwood "moron nursemaids" and the twelve in the control group who had simply remained in the Davenport orphanage. The results were startling: every one of the thirteen led utterly normal lives and had normal IQs, whereas those who had remained in the orphanage led unbelievably desolate and impoverished lives with below- normal IQs. The contrast could not have been greater. And Skeels was lucky enough to trace every single individual in the study, and in keeping with the discourse of the new age, he reported what happened to each *individual*.[44]

Skeels's study turned out to be one of an entire family of similar studies of compensatory preschool education done during the late 1950s and early 1960s. The yeasty enthusiasm of these studies, whose authors wished and claimed that the IQs of disadvantaged children could be raised, led to Head Start.[45] This in turn led to an important critique by University of California Professor of Education Arthur R. Jensen on the very idea of compensatory education in 1969's *Harvard Educational Review*.

Jensen's article has become the subject of all manner of attack. He wrote that compensatory education was a failure, and most likely African Americans as a group or race had lower average IQs than whites. The late 1960s, with the racial and political ferment of the time, was not the most politic time for this statement. In his famous 120-page article, Jensen mainly reviewed various studies of mental testing since Binet. His overall conclusions were based on what he considered the acme of mental testing: the work of Burt and his ideas of factor analysis and inborn intelligence. Simply put, factor analysis was the study of the correlation of various tests of mental ability, which in turn became the proof positive, at least to those like Burt and Jensen, that most of a person's intelligence was inborn and not a response to environmental stimuli. After the 1950s, mental testers who believed in the innate and fixed IQ tended to cite group averages, whereas those who did not believe this theory tended to cite individual scores. On this matter, Jensen was a kind of transitional figure. He was willing to concede all manner of arguments about how an individual's IQ could be determined by a variety of nongenetic factors. Actually Jensen was, in print, far more reasonable on many issues of intelligence and less politically retrograde than he was sometimes depicted in the media. He pointed out, quite correctly, that when retarded infants were moved from an impoverishing social environment to an enriching one, their IQs simply came to what they would have been—they were not really boosted beyond genetic limits. But he stuck with his group averages for blacks and whites. Thanks to the political tenor of the times, his argument that there were innate racial differences between blacks and whites simply begged for trouble.[46]

It was not merely Jensen who got into hot water in a political and scientific controversy. It was also Burt, at least posthumously. Burt had been a major architect of applied psychology in Great Britain, the first psychologist ever knighted, and a solid evaluator of education and educational policies. That he stood on the nature side of the nature-nurture controversy was obviated to some extent in the U.K. by the fact that his political and policy sympathies seemed reformist, which was often the case among British scientists. The difficulty came particularly with studies he did of identical twins that seemed to prove beyond a doubt that social class and IQ were related in a causal sense. The details of the controversy are too arcane to delve into here. Suffice it to say that he was discussing groups, not individuals, and, more to the point, a good deal of evidence and argumentation arose after his death that he had fabricated his data. This seems to have been the case. That controversy, which raged on both sides of the Atlantic for several years with, among others, Jensen as a champion of Burt, did little service to the hereditarian cause.[47]

Another technical development that undercut the position of Jensen, Burt, and others had to do with *g*, that by-now iconic representation of intelligence. After the Second World War, psychologists Wechsler and Thurstone showed that, at best, Spearman's theory of *g* was woefully oversimplified. Wechsler developed mental tests that not only showed the kind of general intercorrelations among many factors that Spearman had unearthed, but also intercorrelations among the many subtests of abilities his tests evaluated. The question became for Wechsler's work, how to interpret the general as opposed to the many more finite results and correlations and to determine what significance each held. No matter what one concluded, *g* as a unitary factor called intelligence seemed to be a historical artifact that was useful, perhaps, for historians of science but not for real scientists. Thurstone's work was even more upsetting to a hereditarian interpretation of racial mental testing. Thurstone was much more interested in the clustering of scores in the subtests, which led him to conclude that there were more factors in intelligence than *g* could possibly represent. He argued that there were at least seven clusters of primary mental abilities, each relatively independent of the others, and, at best, *g* could only represent a general average, which was clearly not of major importance.[48] In other words, changes in science played an important role in changes of attitudes about a racial hierarchy of inferior and superior races.

A last gasp, at least in our own time, came in 1994 with *The Bell Curve* by the late Harvard psychologist Richard Herrnstein and the conservative social policy critic Charles Murray. The authors rehearsed what were by now ancient arguments in the conservative litany of IQ discourse: First, the tests measured something called *g* or a trait that one could consider innate intelligence. Second, there was about a fifteen-point gap in IQ points between whites and blacks. And third, since the science of demography suggested that as a rule higher-status families had fewer children than lower-status families, and since lower-status families had more children of lower IQ than those of the higher-status families, there was what the authors dubbed a dysgenic fact at work in the population, meaning that the lowliest of the low would reproduce at a much higher rate than the most talented, meaning in turn that within a few generations the low life would swamp the high life. The authors' remedy was to construct "reservations" for the less gifted and not allow them to circulate in civil society.[49]

As *New York Times* columnist Frank Rich pointed out in his strong attack, "The Sell Curve," on October 23, 1994, just a couple of weeks after the book was published, the main motive of authors and publisher must have been to sell books, for there were no new arguments or evidence whatsoever in the book, nothing that had not been knocked down, like the target duck at the state fair booth, over and over again, pointlessly—except to the latest new player. He accused the authors of having revived the racist arguments of William Shockley, inventor of the transistor who famously argued in the 1960s that the lower income and higher crime rates of blacks over whites could be explained by the fifteen-point difference in average IQ scores of whites and blacks. And the authors "imagine a polarized, totalitarian "Futureworld" in which the IQ elite would confine the poor and the dumb to high-tech reservations. This and other aspects of the book might make the book, Rich concluded, "the best selling unread book since the last novel by Umberto Eco."[50] Ridicule from Rich aside, *The Bell Curve* seemed to be a nonstarter. After a few weeks of national media controversy and some smoldering fires here and there in various localities, the 845-page book disappeared from the bookstores.

Where did the scientific community stand on the issue of race and IQ in the new century? Scientific and popular opinion seemed to be moving away from racist arguments and from any such group determinism whatsoever. Thus in the 2000 U.S. Census, for the first time, Americans could indicate a multiplicity of racial ancestries. It was both an individual choice and a multiple choice, a sign of the times if there ever was one. And the first mapping of the human genome seemed to make hash of the old notions of race, dependent as they were on outward physical appearance and distinctive traits such as skin color or hair texture. Just before the millennium, in 1998, an international group of psychologists made an important statement in, *The Rising Curve: Long-Term Gains and Related Measures*, a complex book sponsored and published by the prestigious American Psychological Association. As Ulrich Neisser, the volume's editor, pointed out, all three of the claims by Herrnstein and Murray, and others who agreed with them beforehand, were simply dead wrong—wrong as science, in the first instance. One of the most startling discoveries in the last part of the twentieth century is what came to be known as the "Flynn effect," named for New Zealand political scientist James Flynn, who first discovered it. Flynn noted that in every generation IQ scores have been rising for every group in all the advanced countries in the world. "Performance on broad-spectrum tests of intelligence has been going up about 3 IQ points per decade ever since testing began," Neisser declared.

So what did the Flynn effect mean for the three arguments? Consider the existence of *g*. The Flynn effect either did or did not reflect real increases in *g*. If it reflected genuine increases, obviously *g* was influenced by environmental factors, because no genetic process could produce such large changes in such a short period of time. Thus genetic factors in intelligence could not be much of a factor. If, on the other hand, the rise did not reflect real increases, then the tests were obviously flawed and could not be relied upon. Then what about those racial differences or allegations to that effect? In the 1930s and again in the 1980s, the average gap in scores between blacks and whites was one standard deviation, or about fifteen points. Given the Flynn effect, both groups gained some fifteen points during that half century. The gains that African Americans made closed what had been the entire gap. By the 1980s they performed at the level of whites in the 1930s. Was it plausible to assume that differential fertility operated so that the overall genetic potential of the national population would decline in a few generations? This was an old tale that went back to the days—and bromides—of Galton. Yet Flynn's findings, which have been corroborated by many others, suggest that intelligence has been rising, not falling—thus the idea of a "rising curve," not a "bell curve" as Herrnstein and Murray (and their predecessors) would have had it.[51]

In conclusion, what are we to think? Has it always been politics and other centers of authority driving the science, or have they been independent but interacting phenomena? It would seem most defensible (and sensible) to argue that currents of thought and action in the larger culture help shape the contours and structures of science, although they do not determine them in detail. Plainly put, we could say that the larger culture—and politics in particular—raises the issue of racial superiority and inferiority, if for no other reason than that we live in a multiracial nation and world. Yet matters are not so simple. There is a pattern of politics and culture raising the issue of whether science proves there are superior and inferior races, and science, whatever it might mean at any given time and place, providing an answer. Science matters. That is manifestly clear. But there is always a larger cultural context. Clearly there were ages in which Americans believed that the most important thing about a person was that person's identity in a group—a gender, class, race, nativity group, and the like. In such epochs, it made sense, at least from some points of view—certainly the political and the cultural—and what passed for science of the time, to argue that an individual was a member of a group, and that all analysis and proscription should start from that point. Today we seem to live in a very different world, an age that began, without our quite noticing it, in the 1950s, in which there was no longer a relationship between a larger whole and its parts; there were simply as many or as few individuals or parts as one wished to imagine. It was *individual choice*. With that has come many difficulties in our public life, but apparently, as is the case with all cultural shifts from one age to another, there are also some changes that can cause at least a modicum of hope for a better world. If scientific racism truly no longer makes sense in our culture—a brave thought and perhaps a naïve one at that—then perhaps we live in better times than our parents and ancestors. It is probably too much to ascribe the 2008 presidential election as a harbinger of a better, more wholesome climate of opinion among and between the many peoples that inhabit our nation and our world. Or perhaps not. Only time will tell.

Notes

1. Madison Grant, *The Passing of the Great Race, or the Racial Basis of European History* (New York: Charles Scribner's Sons, 1916).
2. Grant, *The Passing of the Great Race* (see n. 1). See also the interesting new biography of Grant, Jonathan Peter Spiro, *Defending the Master Race: Conservation, Eugenics, and the Legacy of Madison Grant* (Burlington, VT: University of Vermont Press, 2007) passim.
3. Comer Vann Woodward, *The Origins of the New South, 1877–1913*, rev. ed. (1951; rev. ed., Baton Rouge: Louisiana State University Press, 1971); Howard Rabinowitz, *Race Relations in the Urban South, 1865–1900* (New York: Oxford University Press, 1976); George M. Fredrickson, *The Black Image in the White Mind: The Debate on Afro-American Character and Destiny, 1817–1914* (New York: Harper and Row, 1971); Joel Williamson, *The Crucible of Race: Black-White Relations in the American South Since Emancipation* (New York: Oxford University Press, 1984); and Michael McGerr, *A Fierce Discontent: The Rise and Fall of the Progressive Movement in America, 1870–1920* (New York: Free Press, 2003), passim, 182–220.
4. John Higham, *Strangers in the Land: Patterns of American Nativism, 1860–1925* (New Brunswick, NJ: Rutgers University Press, 1949) remains the definitive (and elegant) work on the subject.
5. William McDougall, *Is America Safe for Democracy? His Lectures Given at the Lowell Institute* (New York: Charles Scribner's Sons, 1921) vii, 12, 45–72, passim. See also John Corbin, "American Civilization on the Brink," *New York Times*, June 12, 1921.
6. Cornelia James Cannon, "American Misgivings," *Atlantic Monthly* 129 (1922): 145–57.
7. Lothrop Stoddard, *The Revolt Against Civilization: The Menace of the Under Man* (New York: Charles Scribner's Sons, 1923), 1–2, 30–268.
8. George B. Cutten, "The Reconstruction of Democracy," *School and Society* 16 (1922): 477–89.
9. Carl C. Brigham, *A Study of American Intelligence* (Princeton, NJ: Princeton University Press, 1923), 182–83.
10. E. G. Boring, *A History of Experimental Psychology*, 2nd ed. (1929; 2nd ed., New York: Appleton-Century Crofts, 1950), 480ff. The original paper is Charles Spearman, "'General Intelligence' Objectively Determined and Measured," *American Journal of Psychology* 15 (1904): 72–101, 200–292. There was no comparable journal in Britain until the next year.
11. Brigham, *Study of American Intelligence*, 30–31, 159–76, 182–83, passim (see n. 9).
12. Stoddard, *Revolt Against Civilization*, 1–2 (see n. 7).
13. Carl C. Brigham, "Intelligence Tests of Immigrant Groups," *Psychological Review* 37 (1930): 158–65.
14. Hamilton Cravens, *The Triumph of Evolution: The Heredity-Environment Controversy, 1900–1941* (Baltimore, MD: Johns Hopkins University Press, 1988 [1978]), 224–65; Dale Stout and Sue Smart, "E. G. Boring's Review of Brigham's *A Study of American Intelligence*: A Case-Study in the Politics of Reviews," *Social Studies of Science* 21 (1991): 133–42; F. N. Freeman, "An Evaluation of American

Intelligence," *The School Review* 31 (1923): 627–28; A. J. Snow, "Review of *A Study of American Intelligence*," *American Journal of Psychology* 34 (1923): 304–7; Kimball Young, "Review of *A Study of American Intelligence*," *Science* 57 (1923): 666–70; E. G. Boring, "Facts and Fancies of Immigration," *New Republic* 34 (1923): 245–46; Franz Boas, "The Problem of the American Negro," *Yale Review* 10 (1920–1921): 386–95; Boas, "Fallacies of Racial Inferiority," *Current History* 25 (1926–1927): 676–82; and Margaret Mead, "The Methodology of Racial Testing: Its Significance for Sociology," *American Journal of Sociology* 31 (1925–1926): 657–67.

15. Walter Lippmann, "The Mental Age of Americans," *New Republic* 32 (1922): 213–15; Lippmann, "The Mystery of the 'A' Men," *New Republic* 32:246–48; Lippmann, "The Reliability of Intelligence Tests," *New Republic* 32:275; Lippmann, "The Abuse of the Tests," *New Republic* 32:297–98; Lippmann, "Tests of Hereditary Intelligence," *New Republic* 32:329–30; Lippmann, "A Future for the Tests," *New Republic* 33 (1922–1923): 9–11; Lippmann, "The Great Confession," *New Republic* 33:144–45; and Lippmann, "A Defense of Education," *Century Magazine* 106 (1923): 95–103.
16. See Cravens, *Triumph of Evolution*, 224–65 (see n. 14).
17. George W. Stocking, Jr., "The Scientific Reaction Against Cultural Anthropology, 1917–1920," in *Race, Culture, and Evolution: Essays in the History of Anthropology* (New York: Free Press, 1968), 270–307.
18. Franz Boas, "Heredity in Head Form," *American Anthropologist*, n.s., 5 (1903): 530–38.
19. George W. Stocking, Jr., "The Critique of Racial Formalism," in *Race, Culture, and Evolution*, 162–94 (see n. 17) ably covers Boas's work for the United States Immigration Commission and its context.
20. Franz Boas, "Race Problems in America," *Science*, n.s., 29 (1909): 839–49.
21. Franz Boas and Helene M. Boas, "The Head-Forms of the Italians as Influenced by Heredity and Environment," *American Anthropologist*, n.s., 15 (1913): 163–88; Boas, "The Influence of Environment upon Development," *Proceedings of the National Academy of Sciences of the United States of America* 6 (1920): 489–93; and Boas, "The Methods of Ethnology," *American Anthropologist*, n.s., 22 (1920): 311–21.
22. George W. Stocking, Jr., "Franz Boas and the Culture Concept," in *Race, Culture, and Evolution*, 195–233 (see n. 17); and Cravens, *Triumph of Evolution*, 89–120 (see n. 14).
23. Maurice B. Hexter and Abraham Myerson, "13:77 Versus 12:05: A Study in Probable Error," *Mental Hygiene* 8 (1924): 83.
24. Mead, "Methodology of Racial Testing," 657–67 (see n. 14).
25. Alfred M. Tozzer, *Social Origins and Social Continuities* (New York: Macmillan, 1925) 6, 14, 41–42, 53, 62–71, 85; Melville J. Herskovits, "Brains and the Immigrant," *Nation* 120 (1925): 139–41; A. L. Kroeber, *Anthropology* (New York: Harcourt, Brace, and Co., 1923), 75–79; Wilson D. Wallis, "Race and Culture," *Scientific Monthly* 23 (1926): 313–21; and Franz Boas, "Fallacies of Racial Inferiority," *Current History* 25 (1926–1927): 676–82.
26. Boring, *History of Experimental Psychology*, 481 (see n. 10).
27. See Cravens, *Triumph of Evolution*, 235–41 (see n. 14).

28. Among standard works that deal with the Ku Klux Klan after 1900 are David M. Chalmers, *Hooded Americanism: The First Century of the Ku Klux Klan, 1865–1965* (Garden City, NY: Doubleday, 1965); Kenneth T. Jackson, *The Ku Klux Klan in the City, 1915–1930* (New York: Oxford University Press, 1967); Nancy MacLean, *Behind the Mask of Chivalry: The Making of the Second Ku Klux Klan* (New York: Oxford University Press, 1994); Osha Gray Davidson, *The Best of Enemies: Race and Redemption in the New South* (Chapel Hill: University of North Carolina Press, 2007); Leonard J. Moore, *Citizen Klansman: The Ku Klux Klan in Indiana, 1921–1928* (Chapel Hill: University of North Carolina Press, 1991); and Charles C. Alexander, *The Ku Klux Klan in the Southwest* (Lexington: University of Kentucky Press, 1965).
29. The classic statements of this perspective include Margaret Mead, *Coming of Age in Samoa* (New York: William Morrow, 1928); Ruth Benedict, *Patterns of Culture* (Boston: Houghton Mifflin, 1936); Benedict, *Race: Science and Politics* (New York: Viking, 1940); and Theodosius Dobzhansky and M. F. Ashley Montagu, "Natural Selection and the Mental Capacities of Mankind," *Science*, n.s., 105 (1947): 588–90.
30. See Herbert Spencer Jennings, *The Biological Basis of Human Nature* (New York: W. W. Norton, 1930); Thomas Hunt Morgan, *The Scientific Basis of Evolution* (New York: W. W. Norton, 1932); Lancelot Hogben, *Nature and Nurture* (New York: W. W. Norton, 1933); Julian Huxley, *Evolution: The Modern Synthesis* (New York: Harper and Row, 1943); Theodosius Dobzhansky, *Genetics and the Origins of Species* (New York: Columbia University Press, 1937); Dobzhansky and Leslie Clarence Dunn, *Heredity, Race, and Society* (New York: Penguin, 1947); Dobzhansky, *Mankind Evolving: The Evolution of the Human Species* (New Haven, CT: Yale University Press, 1962); and Dobzhansky, "Anthropology and the Natural Sciences: The Problem of Human Evolution," *Current Anthropology* 4 (1963): 146–48. In general, see Cravens, *Triumph of Evolution*, 157–90 (see n. 14).
31. Otto Klineberg, *Negro Intelligence and Selective Migration* (New York: Columbia University Press, 1935), 59.
32. John Dollard, *Caste and Class in a Southern Town* (New Haven, CT: Yale University Press, 1937) passim; and Gunnar Myrdal, *An American Dilemma: The Negro Problem and American Democracy* (New York: Harper and Brothers, 1944), passim.
33. Howard N. Rabinowitz, *Race Relations in the Urban South, 1865–1890* (New York: Oxford University Press, 1978) corrects in some important respects Woodward's classic account, *Origins of the New South*, especially chapters 12–14 (see n. 3). For a more up-to-date account, see Michael McGerr, *A Fierce Discontent: The Rise and Fall of the Progressive Movement in America, 1870–1920* (New York: Free Press, 2003), 182–218.
34. John Higham, *Strangers in the Land* (see n. 4) is the classic account of American nativism and the immigration restriction movement.
35. Peter Novick, *The Holocaust in American Life* (Boston: Houghton Mifflin, 1999) offers a fascinating and stimulating discussion of its subject.
36. Harold M. Skeels, Ruth Updegraff, Beth L. Wellman, and Harold M. Williams, *A Study of Environmental Stimulation: An Orphanage Preschool Project*, University of Iowa Studies, Studies in Child Welfare, vol. 15, no. 4 (Iowa City: The University,

1938), passim. See also Hamilton Cravens, *Before Head Start: The Iowa Station and America's Children* (Chapel Hill: University of North Carolina Press, 1993).

37. Marie Skodak, *Children in Foster Homes: A Study of Mental Development*, University of Iowa Studies, Studies in Child Development, vol. 16, no. 1 (Iowa City: The University, 1939), passim.
38. Harold M. Skeels and Harold Dye, "A Study of the Effects of Differential Stimulation on Mentally Retarded Children," *Proceedings of the American Association of Mental Deficiency* 44 (1939): 114–36; and Skeels, "A Study of the Effects of Differential Stimulation on Mentally Retarded Children: A Follow-Up Report," *American Journal of Mental Deficiency* 46 (1942): 340–50.
39. See Cravens, *Before Head Start*, especially 185–216 (see n. 36), in which these matters are covered in some detail.
40. Gunnar Myrdal, with the assistance of Richard Sterner and Arnold Rose, *An American Dilemma: The Negro Problem and Modern Democracy* (New York: Harper and Brothers, 1944). For context, see Walter A. Jackson, *Gunnar Myrdal and America's Conscience: Social Engineering and Racial Liberalism, 1938–1987* (Chapel Hill: University of North Carolina Press, 1990).
41. See Hamilton Cravens, "American Social Science and The Invention of Affirmative Action, 1920s–1970s," in *The Social Sciences Go to Washington: The Politics of Knowledge in the Postmodern Age*, ed. Hamilton Cravens, 23–31 (New Brunswick, NJ: Rutgers University Press, 2003). See also Diana Selig, "The Whole Child: Social Science and Race at the White House Conference of 1930," in *When Science Encounters the Child: Education, Parenting, and Child Welfare in 20th Century America*, ed. Barbara Beatty, Emily D. Cahan, and Julia Grant, 136–56 (New York: Teachers College Press, 2007).
42. Kenneth B. Clark, "The Social Scientist as an Expert Witness in Civil Rights Litigation," *Social Problems* 1 (June 1953): 5–10.
43. Kenneth B. Clark and Mamie P. Clark, "Emotional Factors in Racial Identification and Preference in Negro Children," *Journal of Negro Education* 19 (1950): 341–50. Ben Keppel, *The Work of Democracy: Ralph Bunche, Kenneth B. Clark, Lorraine Hansberry, and the Cultural Politics of Race* (Cambridge, MA: Harvard University Press, 1995), 97–176, offers an excellent point of discussion.
44. Harold M. Skeels, *Adult Status of Children with Contrasting Early Life Experiences: A Follow-Up Study*, Monographs of the Society for Research in Child Development, vol. 31, no. 3 (Chicago: University of Chicago Press for the Society for Research in Child Development, 1966).
45. For a useful history of this subject, see Maris Vinovskis, *The Birth of Head Start: Preschool Education Policies in the Kennedy and Johnson Administrations* (Chicago: University of Chicago Press, 2005).
46. Arthur R. Jensen, "How Much Can We Boost IQ and Scholastic Achievement?" *Harvard Educational Review* 39 (1969): 1–123. Jensen triggered an enormous controversy, which can be followed in Stephen Jay Gould, *The Mismeasure of Man* (New York: W. W. Norton, 1981), to which Jensen answered in Jensen, "The Debunking of Scientific Fossils and Straw Persons," *Contemporary Education Review* 1 (1982): 121–35.
47. L. S. Hearnshaw, *Cyril Burt: Psychologist* (Ithaca, NY: Cornell University Press, 1979), passim. This is a good biography. For a brilliant and highly technical

example of the difficulties of defending Burt's most controversial and likely faked studies, see D. D. Dorfman, "The Cyril Burt Question: New Findings," *Science* 201 (1978): 1177–86.

48. Raymond E. Fancher, *The Intelligence Men: Makers of the IQ Controversy* (New York: W. W. Norton, 1985), 149–61.
49. Richard Herrnstein and Charles Murray, *The Bell Curve* (New York: Free Press, 1994), passim.
50. Frank Rich, "The Sell Curve," *New York Times*, October 23, 1994.
51. Ulric Neisser, ed., *The Rising Curve: Long-Term Gains in IQ and Related Measures* (Washington, DC: American Psychological Association, 1998), 4.

Robert Coles and the Political Culture of the Second Reconstruction[1]

Ben Keppel

There is no keeping up with history, but I hope the record set forth here will be of some worth to future historians who find themselves interested in how a few Americans lived through a critical time in their lives and this country's development as a democracy.

—Robert Coles, *Children of Crisis: A Study of Courage and Fear* (1967)[2]

Social Science and Common Knowledge

Nearly fifty years ago, in 1960, Robert Coles, a Harvard-trained physician and psychiatrist, newly married to fellow New Englander Jane Hallowell, returned to the South. Only a few years earlier, as a captain in the United States Air Force, he had been stationed, not by choice, at Keesler Air Force Base in Biloxi, Mississippi, where he had administered a neuropsychiatric hospital. This time, Robert and Jane Coles, with the moral and financial support of their families, returned as a team to study the lives of children at the center of a region undergoing desegregation.[3] The result of that second journey south, and the primary focus of this essay, is volume one of the Children of Crisis series, published in five volumes between 1967 and 1977.[4]

In February 1972 the publication of volumes two and three, published simultaneously at the end of 1971, were marked by a cover story on Coles in *Time* magazine titled "America's Forgotten Children."[5] Ruth Mehrtens Galvin, editor of *Time*'s Behavior section, researcher Virginia Adams, and reporter Nancy Newman honored Coles as "the most influential living psychiatrist in the [United States]," not because of a breakthrough in theory but for a widely admired public experiment in which his subjects were also his intellectual collaborators. Coles was worthy of this special public attention because he was a genuine expert who was also deeply suspicious of the dogmas that can be induced by too reflexively observing the otherwise legitimate routines required to become an expert. Most of all, Coles was recognized for his work in testing Americans' citizenship on the front lines of the nation's struggles with poverty and racism.

By rising above the set prejudices of both liberals and conservatives, Coles helps depolarize a divided society. He has performed one of the most difficult and important feats of all: to criticize America and yet to love it, to lament the nation's weaknesses . . . while continuing to cherish its strengths. *Most important, he avoids the sterile dogma of social science and speaks, unashamedly, from the heart.*[6]

These words convey clearly how the authors interpreted Coles; they also suggest his importance, especially to the liberal wing of the establishment of his day. Stepping back a bit from what is said here about Coles himself, this rendition is also historically useful for orienting us to the world of Coles's public life and the politics and political culture of the Second Reconstruction.

The Second Reconstruction (to which the civil rights movement was catalytic) has the *Brown* decision in 1954 as its opening boundary and the election of Ronald Reagan to the presidency in 1980 as its closing scene. American political culture was fairly stable in the 1950s, and its ideological distinctions were fairly fuzzy. This was the result of a welcome prosperity, a campaign of fairly explicit cultural and political repression drawing upon a unifying threat from abroad.

Over time, the civil rights movement, whose leaders emphasized the contradiction between segregation and other forms of racial oppression at home and the avowed national outrage over iron curtains and captive nations abroad, pierced this enforced consensus. As dissent became more acceptable, the confluence of increased affluence and dramatically rising educational levels was joined with the emergence of new controversies over race, gender equality, and a land war in Southeast Asia that polarized the public. Although polarization is not generally regarded as a positive development, it accomplished some salutary ends: the political opinions Americans held in this era, as measured by public opinion polling, became somewhat more sophisticated, clearly articulated, and intellectually consistent.[7]

Affluence and education also had important consequences for the relationship of social science knowledge to common knowledge. When we look back on the early history of social science, it is very clear that social science often took the role of ratifying reigning ideas about the ranking of races and peoples, which are now properly recognized as racist.[8] In the period under discussion, as the membership of the social sciences grew and diversified socially and intellectually, the range of opinion about what social science recommended on public questions also grew. Social scientists have always been engaged in "political work"—that is, work bearing pretty directly on controversial public questions. Beginning most famously with *Brown v. Board of Education*, however, some leading social scientists were endorsing positions on social and political questions that were often at least somewhat ahead of what an electoral majority in the United States might endorse. As the debate over implementing "desegregation" and "integration" widened and divided the public, the experts not only became targets of political invective but they also became divided about what their findings meant for public policy. In this situation, the line between being an "expert" and the lower rank of "political advocate" was effectively erased, posing a new set of challenges, especially for practitioners who wished to address the general public.[9]

The cover story on Coles, which came three years after *Time* had inaugurated the Behavior section in its magazine, is evidence of another related development in the expanding public visibility of the social sciences: how ideas from psychology and sociology became, in the words of sociologist Hylan Lewis, part of the "verbal technology of the times."[10] In the political culture of the Second Reconstruction, Lewis argued, the pressure to be politically "relevant" (or at least fashionable) and to catch the wave of intermittent public interest in certain social problems led to a blurring between the "legitimately scientific" and "faddish and transient popular shorthand."[11]

It was in this context that Coles worked. In the late 1960s Coles was becoming an especially ubiquitous figure. The first volume of Children of Crisis was but one of Coles's writings available to the attentive public; by this time Coles had already written about his work in professional journals and journals of liberal opinion, especially the *New Republic*, for several years. As Galvin sought to get oriented to her new journalistic beat, she kept hearing about Coles and encountering his many writings. Galvin and Adams, her one colleague in the first years of the Behavior section, saw in Coles a kindred spirit. "He does what we also try to do—to tell what people are really all about and how they come to be the way they are," Adams wrote.[12]

In the world in which Coles conducted his fieldwork (the United States in the 1960s and early 1970s), political leaders, policy makers, and voters were deciding whether a Second Reconstruction would take place or not. As they had during the first Reconstruction, Americans ultimately gave a mixed answer to this question. Between the end of the Second World War and the climactic moments of the Second Reconstruction, public support for government intervention to ensure the general welfare and the proposition of racial equality dramatically rose to high levels and have remained there despite a rightward turn in the nation's politics.[13] Coles sought to promote the process of constructive peaceful change by challenging insufficiently informed generalities, especially about the poor and the powerless, with concrete knowledge about individual lives under stress. As he worked, Coles also came to see himself as arguing against an attitude held by professionals in the social and behavioral sciences who seemed overly preoccupied with typing and classifying human behavior rather than understanding its individuality and variety.

In his life as a participant-observer of and in certain American lives, the economically comfortable, Harvard-educated Coles straddled the boundaries of class and culture. Whether Coles was working in New Mexico or Atlanta, he and his family would always settle in "a working class neighborhood . . . I would always try to get away, try to be with ordinary people as neighbors."[14] In this carefully considered synthesis of life and work, Coles endeavored to act as "an intermediary of sorts, someone who tries to hear and then prompt others to listen."[15] These skills as a diplomat and a go-between served Coles well not only in his work with families enduring various kinds of hardship but also in walking through the ideological minefield created during the 1960s, when debates within social science about the African American family structure and the nature of poverty and the poor spilled into the larger political culture.

I have found it useful to think of Coles as practitioner of "cultural exchange." This idea, represented most famously by the Fulbright fellowship program (named for United States senator J. William Fulbright of Arkansas, who proposed the program as the Second World War was coming to an end),gave institutional substance to an important theme in the public diplomacy of the Cold War years. The Fulbright program was but one of many efforts to increase culture contact and exchange between peoples. The hope was that the exchange of artists and scholars between the United States and other nations would accomplish what professional diplomats by themselves could not: enhance human understanding through thousands of direct personal interactions between people who could build upon the foundation of a shared calling or interest. There were, of course, other motivations at work here: when the Fulbright program sent a Lincoln scholar to a lectureship in Eastern Europe or the State Department sent African American jazz musicians to the capitals of "non-aligned nations" in the "developing world," it was an exercise of American "soft power" in a larger ideological struggle. As genuinely idealistic as some of the motivations for exchange indeed were, there is no getting around the fact that this was not an encounter of geopolitical equals. Americans visited other countries with strong dollars in their pockets and, whatever their precise intellectual mission, fortified by the conviction that they were visitors from the most powerful center of that universe. Those who came to the United States came from either once-great imperial nations now demoted to more peripheral roles in world affairs or new nations struggling to find their way in a postcolonial world. Either way, "center" or "periphery" remained stable conceptual frames applied to a rapidly changing map.

In a similar way, the Coles cultural exchange program was fraught with potential problems. He was a man with resources: education (especially medical expertise), money, and contacts. His subjects possessed few, if any, of these resources to exchange. Coles was acutely aware of this as he wrote about the lives of the hard-pressed and the downright poor. Beyond providing direct assistance to individual families, Coles sought to even out the exchange by turning over a larger and larger part of each volume of Children of Crisis to the direct testimony of these "untrained" individuals. They share the interpretive stage with him to such an extent that Coles sometimes seems to disappear from view. Coles's subjects became his collaborators, giving evidence that many people without formal education nonetheless understood their situations quite well—to an "advanced degree," in fact—and would be up to the challenge of taking a fuller part in society if given the opportunity.

As pieces of psychologically detailed social commentary aimed at socially concerned experts and citizens such as Coles himself, the five volumes of Children of Crisis were a success. They are, however, also a kind of failure. Coles's experiment in cultural exchange was conceived at a time when liberals were seeking ways to connect not only with the very poor of all races but also with the working-class and middle-class families for whom the populist rhetoric of George Wallace resonated strongly. The defeat of George McGovern in the presidential election of 1972 was a powerful symbolic event in this process, suggesting the heretofore unappreciated difficulty of the road ahead for a fuller social and political reconstruction and the unexpected ways in which class inequality, economic stagnation, and feelings of cultural marginalization might move voters quite far to the right—and away from their presumed economic and social interests. In this sense, the Children of Crisis series is a kind of relic from an experiment that failed.

"White Knight"

When newlyweds Robert and Jane Coles embarked on their Southern research in 1960, they had relatively little knowledge of the complexity of the work they were undertaking. In 1960 a black social scientist could not have conducted this research; no less a figure than psychologist Kenneth

B. Clark (who had conducted his famous doll research in the South and border South) explained to Coles how the continuing fact of segregation might well fundamentally hinder his work and cloud his findings:

> *If you were a Negro and poor, you'd be hearing different words from those . . . children, writing up a different research project; and you sure wouldn't be thinking of (not in your wildest dreams!) doing interviews with white children and their parents—or with the school teachers. . . . I couldn't do [your] project either—even in Negro homes . . . I'm a Northerner and I've lived a comfortable protected life; it would be hard for me to leave it for that world of danger and violence.*[16]

Looking back upon these years from the distance of thirty years, Coles candidly concluded that being white gave him an advantage as a social investigator. In a society in which white privilege was normative, whiteness coupled with education and social standing had "served to protect me . . . I was a white knight who needed have no fear, and who thought . . . he could go anywhere and talk to anyone."[17] Such confidence, Coles argued, however admirable, is also a privilege, an entitlement that comes with a certain kind of life that, to begin with, was lived on the white side of the color line. Coles confessed that "I cringe today at my naiveté and my self-assurance, and maybe my unknowing . . . arrogance—even as I doubt that without that psychological, never mind racial and social background, I could have gotten even to first base on either side of those railroad tracks I visited in those beleaguered Southern Cities of the early 1960s."[18]

The first observation that Coles offered readers of *Children of Crisis: A Study of Courage and Fear* was that in his first visit to the South as an air force officer, he had been unaware of the way in which the segregated social system supported him: "I recall how easily I slipped into its very distinctive life and how pleasant I found that life to be as a white, middle-class professional man, and so I fit easily into . . . Southern society. Only gradually did I begin to notice the injustice at hand, and as a consequence take up my particular effort against it."[19]

Over the course of his work as a participant-observer, Coles became deeply skeptical about the value of social science methods and objectives. This attitude is not as evident in volume one of the series, but it is an explicit theme in succeeding ones. In volume two, Coles admits to being "methodologically untrained" but suggests that this is an asset rather than a liability. The student schooled in the literary imagination looks closely to the particular, seeing the person, not just the social category into which some other social scientist observer has placed them.[20] The expert and the advocate's "hunger after certainty . . . 'orientations' and 'conceptual frameworks'" is often driven by an unexamined need to "do things *for* people, *with* them and even, alas, *to* them. Therefore, we see what we want to see, and find important." From literature rather than social science (specifically the works of Flannery O'Connor, William Faulkner, and James Agee), Coles has learned that "there are many truths—so many that no one mind or viewpoint or discipline or profession can possibly encompass and comprehend them all, nor do justice to them in words, even intricate and specialized ones or 'neutral' ones."[21] For Coles and a growing number of colleagues, the people he studied must be seen "as citizens of a nation, as members of a given society, and *particular* members at that, not merely members of an Oedipal family."[22]

In volume one of the Children of Crisis series, *Children of Crisis: A Study of Courage and Fear,* Coles concentrated on children as focal points in a larger community. In this context, it is important to remember some other forces at work in American society. The combination of a "baby boom" and the impact of television, news magazines, and paperback books spread new knowledge about child health and child development into every part of American society, including "cabins and tenements far removed from our national life."[23] Although Coles gave children important attention as principal actors and symbols, he was equally attentive to the lives and ideas of their parents and their teachers. What resulted was an unusually detailed time-lapse photograph of how human beings take in, live with, and interpret the culture around them. Although Coles's other writings from the early to mid-1960s evinced a skepticism about the operational assumptions of social science, Coles did not give major attention to a critique until volume two, *Migrants, Sharecroppers and Mountaineers,* published in 1971. In volume one, Coles discussed social science primarily to situate what he was doing as a clinician in relationship to the social scientist:

> *Social scientists have the . . . very necessary job of documenting exactly what "outside" world it is—the time, the place, the culture and the society—that a particular child finds at birth and learns about as he grows. The clinician has always been interested in what goes on "inside" others—he himself being part of their "outside" . . . Presumably, he is as qualified as anyone to go back and forth, to see how the two worlds . . . connect, blend, engage.*[24]

Writing four years before the publication of *Children of Crisis: A Study of Courage and Fear,* Coles described the method that he and Jane Coles followed in pursuing this work:

> *Our methodology . . . was to go from home to home . . . to talk, to listen, to play, to draw. We went South to find out about the psychological problems in desegregation among young and old, Negro and white . . . Going into our third year, we still seem to learn, to correct our vision, to feel newly aware or lately wrong. I shudder now at some of my first impressions about how Negroes feel about segregation, or whites, for that matter. Yet isn't this always the demand of our work, its need for time to gauge acutely what time and its events have "done" to people.*[25]

In this process of discovery, Coles did not exclude a very searching analysis of his own behavior in the midst of this social stress. He candidly described his own shifts in mood and commitment over the course of a series of meetings he had with members of the Student Nonviolent Coordinating Committee and members of the Nation of Islam (whom Coles simply referred to as "muslims"). These reactions were not treated as "expert" knowledge but rather as something that connects Coles to his fellow Americans.[26]

A careful review of Coles's works confirms that, in addition to his debt to the literary figures listed earlier, Coles was especially steeped in the psychologically centered social science of the 1940s and 1950s, especially cultural anthropology. Coles's early reckoning with the theory of what racism actually did to people was informed by Abram Kardiner and Lionel Oversey's *Mark of Oppression*, which emphasized the damage caused to the Negro personality by white racism. In his first outings, Coles applied his new knowledge with all the zeal of the new convert. Jane Coles criticized his early propensity for always "characterizing the people we see! Why don't you let them *be*. Why don't you pay attention to each person in each family that we visit, and stop trying to lump them together."[27]

From the anthropologists Allison Davis and John Dollard and especially Oscar Lewis, Coles had learned the merit of letting "the people speak without interruption," a mode of presentation that would characterize future volumes in the Children of Crisis series.[28] Coles developed a warm professional friendship with Lewis and credits Lewis with introducing him to the tape recorder as a research tool.[29]

Although Coles's chosen medium for examining how children blend their external and internal worlds was drawings rather than dolls, Kenneth B. and Mamie Clark's emblematic doll studies have a special importance beyond Coles's personal relationship to Kenneth B. Clark. In an important sense, Coles was elaborating a line of inquiry with which the Clarks are very closely associated; he was measuring how core social values were reflected in children's choices about how to represent the world they see and experience. Coles's fieldwork also partially confirmed the Clarks' findings that children have generally developed extremely negative associations with the colors brown and black. At the same time, a close comparison of Coles's and the Clarks' work reveals important differences. The Clarks' work with children focused rather narrowly on reactions that are manifest, as children are asked to relate themselves (through the act of choosing one doll over another, for instance) to the outside world. From this, the Clarks assumed an internalization of negative social and cultural messages from outside. The reader does not learn the life history of each child or meet his or her parents. The Clarks were concerned with documenting something very important but narrower in scope: that very negative social and cultural messages had been received and, at the very least, understood. Coles was trying to assess what the child actually does with those messages over the course of a childhood—the particular, individual, and unique meanings that he or she applies to them.

When he studied the drawings of the many children Coles and his wife visited over hundreds of hours, Coles was looking for clues to how children made sense of the culture around them. The children's drawings were full of "energetic symbolization" about the world and its meanings and messages.[30] It took six months, for instance, for six-year-old Ruby Bridges to use her brown or black crayons to represent anything other than dirt. In the work children drew for Coles, size was as important as color as a way of representing the differences in status and power created by segregation. According to Coles, Ruby's representations of white people were "larger and more life-like. Negroes were smaller, their bodies less intact."[31] In Ruby's drawings, however, African Americans were given larger ears, perhaps because they needed them to compensate for their lack of power in other areas.[32]

If capturing human feeling and experience requires listening to many different voices telling many different stories, Coles also understood that one color or one human feature can hold more than one symbolic meaning.[33] Coles's analysis of the drawings by African American children led to a negative conclusion about the symbolic meaning of brown and black. "Every Negro child I know has had to take notice in some way of what skin color signifies in our society. If they do not easily . . . talk about it, their drawings surely indicate that the subject is on their minds. Like Ruby, they have trouble using brown and black crayons."[34] At the same time, when Ruby drew her grandfather in his pastoral setting, Coles noted that "brown and black were used appropriately and freely" under the cloak of protection offered by the memory of her grandfather, including to represent a very large brown man who was very much intact. ("Momma . . . says her daddy is the strongest man you can find," she said.) For Ruby, her grandfather's Delta farm was a refuge where it was safe to be yourself. ("If things gets real bad [sic] we can always go there," she said.)[35]

Coles's long exposure to black children and their families enabled him to see a fuller picture of how strength and weakness often coexist in one person or in one family. Coles agreed with the findings of the famous "Moynihan Report," written in 1965 by his friend and colleague Daniel Patrick Moynihan, on the special problems facing the black family. At the same time, Coles found Moynihan's presentation problematic and the controversy generated by the report unproductive. In a letter to David Riesman, Coles expressed frustration at how the lines of debate over the report had become drawn: "Why we have to come out for or against something is perhaps a more important problem or issue . . . I think I know what [Pat] was trying to do, and I am for it, but I also think that some of things that he says have to be much more carefully thought about than he has yet done."[36] At the same time, Coles found Moynihan's awareness that the dynamics of class were compounding the injustices of race to be an important contribution to the public dialogue. Finally, Moynihan's sincerity of motive (perhaps emphasized for Coles by the rhetorical overkill of some of the attacks on his friend) ultimately caused Coles to call Moynihan "just about the most courageous man in America today" for his outspokenness.[37]

In *Children of Crisis: A Study of Courage and Fear*, Coles provided an alternate model for addressing both the damage and the endurance embedded in the lives of many African Americans. "Fear, hate, hunger and utter segregation from the rest of the community establish an atmosphere that inevitably affects the child and the parent every day and profoundly." At the same time, Coles made clear that the damage to children resides not in some flaw deep in the African American family (he found that "in the first years of infancy and early childhood" African American families, though hard pressed, often successfully provide comfort and protection) "but in those later years when the world's restrictions become decisive antagonists to the boy or girl" who must "come to terms with the distinct fact of the Negro's powerlessness in American society."[38]

The controversy over the Moynihan report was one moment in a longer national conversation about how American society should be reconstructed to eliminate racism and its many consequences for American society. The particular moment created by Moynihan to promote his report is widely remembered as, in Moynihan's words, "a moment lost"—an opportunity lost in unproductive argument between allies (Moynihan and leaders in the civil rights movement such as Martin Luther King, Jr.). However, Moynihan's part in this loss has also been missing from the remembered record.[39] Moynihan was far more a politician ambitious for political advancement than he was a social scientist seeking to make his name by provocatively breaking new ground in a controversial area. Moynihan was neither a racist nor an empty opportunist with no genuine commitment to the issues he championed, including the black family. Among politicians, Moynihan does not seem (to this author) unusually grasping or opportunistic, but he was (at thirty-eight) a young man in a hurry whose report was vague and sloppy because it was quickly written in part, perhaps, to coincide with its author's entry into New York City electoral politics as a favorite candidate of the Johnson White House for major citywide office. Moynihan leaked the report to Robert Novak (the first person to refer in print to the document as the "Moynihan Report") barely one month before primary day in Moynihan's only unsuccessful campaign, to become the Democratic nominee for city council president (a position filled by a citywide vote).[40] This part of the political puzzle has been missing from historical accounts because the report seems to have played a very small role in the campaign or in Moynihan's defeat. The famous controversy took time to build, arriving in 1966 and continuing well into 1967.

In respect to this "Moynihan moment," there is certainly a cautionary lesson here. Addressing complicated social problems, especially those so centrally connected to race and class, in a political culture with shortness of vision on both types of issues, requires special work. The preparations required on the levels of politics and policy require years, not months. The closing and most publicly visible phase of such a process must be focused around a proposal that is concrete, specific, and addressed to one significant and substantive part of an issue that is admittedly much larger and more complex. And even these steps are no guarantee of success.

Moving Moynihan to one side, it is also worth considering some larger and less personal factors in the failure of the debate to generate a positive policy result. Even at the high point of the Great Society, it is far from clear that a large-scale social program of family allowances could have been enacted. Left unaddressed (then and later) was the overarching fact that, beginning in the 1960s, working and earning wages in America was taking place in an increasingly diffuse and competitive global economy in which the United States was but one of many highly developed economies. For African Americans, this was a development with special implications. As historian Earl Lewis has argued, the economic and political shackles formally imposed by segregation were coming off just as the general price of entry into the American middle class was rising steeply.[41]

Potentially powerful historical moments don't come very often, and they can be easily lost. In this case, Moynihan and some of his more inflammatory critics must share responsibility for the fact that this Moynihan moment made it more difficult to discuss the connections between a changing economy and the general stability of American families, and between historical exclusion based on race and continuing exclusion rooted in a changing economy.

Behind the White Line

Despite the fullness of his preparation, Coles was unable to surmount some barriers to research raised by segregation itself. Coles's picture of white Southerners in crisis because they opposed desegregation efforts is not quite as full as that of the black Southerners. Looking back to 1971, Coles, in perhaps his single most important essay on his fieldwork, "The Observer and the Observed," discussed how he wrestled with his own reaction to what he heard and how he conformed to the stereotype held by his white listener of the coldly and silently judgmental professor whose mind was busy filing away this man as a social type. He was prodded by Jane Coles to look again, this time more closely, to see experiences and attitudes that Coles had in common with his interlocutors:

> *"He and his wife are so predictable," I once said to my wife as we were leaving the house—after they had been thoroughly kind to us, fed us . . . delicious fried chicken and okra and cooked tomatoes and mashed potatoes and home-baked bread and pecan pie topped with ice cream. If I was annoyed, my wife was annoyed with me. She sensed my irritation and impatience and at times outrage. She too felt dismayed and saddened by what she heard. But she also felt something else: she felt that in such a home we were in the presence of more than odd or idiosyncratic or unusual people . . . For my wife, "they" were in many respects "us"—if not all of us, then most of "us" . . . She simply would not stop pointing out that if there were differences between us and some of the white hecklers we had met on the street, some of the white people we were getting to know, there were also many things we shared . . . as white people, as Americans, as men and women alive in the second half of the twentieth century. Nor would she allow me to get away with saying that my job was to find out how the black children were getting along.*[42]

In later books—especially *The Buses Rolled* and *The Middle Americans*[43]—Coles gave equal attention to those being pushed to make a change they deeply opposed; the racism is there, tangled up in complicated ways with other emotions that cannot be summarily discounted as window dressing for racism.[44] With respect to the representation of white families mobilized against desegregation in New Orleans in volume one of Children of Crisis, Jane Coles's victory was, at that point, only partial. I have searched carefully and can find only one segregationist counterpart to Ruby Bridges there. Readers may remember Jimmy, Ruby's white classmate, who is clearly an ideological racist, although a sometimes inconsistently practicing one. That he was a classmate at all is a fly in the ointment; Jimmy's parents may not have had the financial means to send their child to a private school. That option would have enabled them to buy their way out of the desegregation process and, perhaps, out of meeting the Coles research team. Jimmy's parents withdrew him when the protests against desegregation began and returned him once they had subsided. We do not hear from them about this decision, and there are no word portraits of this family's individual members comparable to those Coles provides of the Bridges family.[45]

Though segregation may have kept them apart until the time of the study, Jimmy and Ruby's drawings show a shared distorted perception of brown people. When Jimmy, an exacting reproducer of details in other contexts, drew Ruby, Coles declared that "it is almost as if he had suddenly embraced surrealism." For Jimmy, brown was the key symbolic detail for understanding people thus colored. "Either they were in some fashion related to animals, or the color of their skin proved that, if they were human, they were dirty human beings—and dangerous too."[46]

Under questioning from Coles, Jimmy explained that "she's funny. She's not like us, so I can't draw her like my friends. Besides," he added in a moment of gifted observation, "she hides a lot from us."[47] Over the course of the school year with Ruby, Jimmy's drawings changed, reflecting an emergent and at least a partially new consciousness. Jimmy's drawings showed the evolution of another individual's internal response to a changing outside world. "In time [Jimmy] regularly came to see [Ruby] as an individual. Amorphous spots and smudges of brown slowly took form and structure . . . Eventually she gained eyes and well-formed ears."[48]

The segregationist of whom we get the closest view is an adult identified as Mrs. Patterson. Mrs. Patterson did not have the means to send a child to private school to avoid desegregation. Her economic poverty had become matched by an exhaustion and poverty of the spirit. At this point in her life, this drastic change in the social order was beyond her emotional means to accept. "I have enough to do," she told Coles, "just to keep us alive without Niggers coming around. They're lower than a dog in behavior. At least he knows his place and I can keep him clean. You can't do that with them. They're dirty. Have you ever seen the food they eat? They eat pig food and they eat it just like pigs, too."[49]

According to Coles, Mrs. Patterson, the mother of five children between the ages of four and seventeen, was fighting a losing battle of attrition with some dire circumstances. Coles wrote, "She is struggling to manage herself in the face of poverty, ignorance, social isolation . . . and virtual abandonment by her occasionally employed husband."[50] This language and the imagery they bring to mind are another summons to remember the lost opportunity represented by the skewed discussion of the Negro family in the mid- and late 1960s. Poverty, ignorance, and social isolation are not the flaws of one person, people, or race but conditions confronted by many people in many circumstances. As one child told Coles, "The rich folks decide how the poor folks live."[51] This sense of class resentment was shared by many whites, even some like the white New Orleans fireman's wife who challenged Coles:

> *You want to know how my children like going to school with the colored? Why don't you talk to the people in the federal government or the rich people in the Garden District? Ask them what they think they have done to us. You can't really know about plain working people like us, unless you go find out about the big shots. It's their decisions that make us live the way we do.*[52]

Dr. Coles Goes to Washington

Coles's education in the practical politics of translating research findings into an improved society formally began in 1965, the first year of the Great Society's War on Poverty. Coles spoke on behalf of the Mississippi Child Development Group, which was trying to bring the Office of Equal Opportunity's Head Start to Mississippi. By this time he was a heavily credentialed expert with considerable fieldwork behind him.[53] Though he found many aspects of the program that could be improved, he was generally quite pleased. "In my ten years of work in child psychiatry, I have yet to see a program like this one, in the sense that against almost impossible odds children have been taught and also receive the benefits of medical care in a manner and with a thoroughness that is truly extraordinary . . . this program is actually reaching the poor children and doing so in a way that that makes any doctor feel truly impressed."[54]

The prospect of Head Start coming to Mississippi was exactly the kind of outside intervention the state's political establishment feared most. A very strong counterattack was quickly launched by the Mississippi political establishment. A lead editorial in the *Jackson Daily News* saw "a disturbing aura . . . terrifying in its ultimate projection hovering above the Head Start phase of the government's poverty program." The prospect of a program open to small children of both sexes and races reminded these editorial writers of the youth indoctrination programs pursued by the governments in Communist China, Soviet Russia, and Nazi Germany.[55]

This was all part of a political offensive led by Senator John Stennis, which, according to Polly Greenberg, an outspoken leader in the struggle to establish the Child Development Group of Mississippi, began with a threatening phone call from Mississippi governor Paul Johnson to the head of the Office of Equal Opportunity, R. Sargent Shriver, castigating Shriver for funding "mah [sic] political enemies." This was the first step in restoring the balance of power in favor of the traditional political leaders of Mississippi.[56] Coles recounted the consequences this way:

> *A beleaguered [Child Development Group of Mississippi] had to fight . . . a hopeless battle: the state of Mississippi was its avowed enemy, and as for the federal agencies . . . I'll let another mother I heard speak at a meeting make her summary: The government people up there in Washington DC—they're more afraid than we are. We must have gone through a lot down here when the day comes that people representing the United States government are more afraid than we are. It used to be [one year earlier] they'd come down here and tell us we shouldn't be afraid.*[57]

The publication of volume one opened a new Washington-centered phase in Coles's work within the political system. In addition to his essays in leading journals, Coles also regularly testified before Congress between 1966 and 1973, a period characterized by both social breakthrough and political backlash. Between 1966 and 1968, Coles also developed a particularly close friendship with Senator Robert F. Kennedy, who admired *Children of Crisis: A Study of Courage and Fear*. According to Coles, it was Kennedy who urged him to broaden his in-depth research of the poor to include Chicano, Indian, and Eskimo children.[58] In his devotion to listening to all voices and his passion for understanding the people who chose George Wallace over Hubert Humphrey or George McGovern when they voted (if they voted at all), Coles was performing the bridge-building work in social science that Kennedy attempted in electoral politics.[59]

As a recognized force—especially in the liberal neighborhoods of the American establishment—Coles used his contacts with politicians to raise the visibility of his research. Emerging national liberal leaders (and future candidates for president) Walter F. Mondale, George McGovern, and Edward M. Kennedy sought his expertise and lent him political credibility by providing introductions to very short distillations of his fieldwork.[60] In content, the books were lavishly illustrated reminders not only of the cultural ambiance of the 1930s but also that racism, fused to the political economy of the South, continued to configure political necessity in such a way that two presidents central to the American tradition of liberal reform—Lyndon Baynes Johnson and his most important political mentor, Franklin Delano Roosevelt, each in his time, would decide that it would be best step back from pressing for more comprehensive change until the political realities improved.

After Robert Kennedy, Coles's most important friend and ally in the political world was U.S. senator Walter Mondale. The correspondence in the Coles papers extended from 1967 until 1976, when Mondale was elected United States vice president. What began as a correspondence between "Senator Mondale" and "Dr. Coles" in March 1967 became an exchange between "Fritz" and "Bob" by August 1969 and continued in that tone through 1976. In his work as an occasional essayist for the *New Republic*, Coles lauded Mondale's frank advocacy for the poor and the powerless. He followed the publication of his profile "Mondale of Minnesota: Champion of Powerless People" in the December 25, 1971, issue with a warm letter, looking to the role he hoped Mondale would play in the 1972 election campaign:

> *If I were Edmund Muskie and had just received the nomination of the Democratic Party for the Presidency, I would turn to you and ask you to be a running mate. The both of you could go to the working people of this country, the factory workers, the blue collar workers and white collar workers, the small farmers, and raise hell with the way a whole range of things are being done in Washington. There's plenty of populism left in the people of this country, and it is simply there waiting to be approached sensibly and forthrightly.*[61]

Sometimes the necessities of research complicated political involvement. Although Coles accepted an invitation to advise the McGovern campaign in 1972, he did fear that too high profile a role would interfere with his fieldwork among working-class whites in Boston.[62]

From the public platform created by the Children of Crisis series, Coles sought to engage with the public by challenging enduring stereotypes about the poor and drawing a stark contrast between a generous and Olympian rhetoric about the generosity of its government and the miserliness of the daily operational practice. Two years after the passage of the Voting Rights Act, in an essay for the *New York Times Magazine* timed with the publication of *Children of Crisis: A Study of Courage and Fear*, Coles opposed the tyranny of certain all-American dictators with the right of even the very poorest American to live and be treated as a citizen:

> *In Alabama and Mississippi and elsewhere in the South, poor people of both races remain hopelessly divided by years of hate and fear. They are still the victims and the oppressors still the jailers who beat the "uppity" ones, still the Governors and would be Governors who rant and rave, and worst of all, people by the many thousands who are lucky to earn $800 or $900 in a year. School libraries, the vote—they are . . . irrelevant in the lives of migrant farmworkers, tenant farmers or sharecroppers, all of whom live as citizens of this rich, strong and God-fearing Republic.*[63]

Readers of the same article would also receive from a little African American boy drawing a brown school bus (with a face for an engine and eyes for headlights) a child's-eye view of social change:

> *Children are like dreamers: they remember everything and say or draw the most explicitly symbolic things. I have no doubt that, in this instance, a Negro boy was showing me what older Negroes keep telling me in both the South and the North keep telling me: the past was the past. The bus was carrying him away from ladders, trees and death . . . he was moving toward the sun, toward something that would promise life and freedom. . . . [The boy tells Coles,] "The bus can go real fast, so if anyone gives us trouble, we can show him."*[64]

Speaking before a subcommittee hearing titled "Migrant and Seasonal Farmworker Powerlessness" chaired by Mondale, Coles addressed the question of powerlessness by drawing a direct comparison between the migrant laborer and the American space program a little more than a week after the first moon landing. "They live fifty miles from Cape Kennedy. No billions have been spent on them and they are not going any place including the moon. They are just living where they are, though. They do a lot of traveling up and down the Eastern seaboard. They are predominantly children."[65]

In the same hearing, Coles spoke directly to another enduring stereotype of American political culture: the idea that people are poor by choice or out of laziness. "No group of people I have worked with . . . tried harder to work . . . in exchange for the desire to work . . . these workers are kept apart like no others . . . are denied even half-way decent wages, asked to live homeless and vagabond lives, lives of virtual peonage."[66] Coles's testimony was part of a successful larger effort to end chronic hunger in the United States. Coles was one of several advocates whose persistent effort led to a significant expansion of the Food Stamp Program and the school lunch program.[67]

Like so many liberals of the time, Coles wanted to understand the problems faced by white Americans who, in politics and social background, were so different from himself. Specifically, how could white middle- and working-class voters be brought within a coalition supporting not only measures to secure equality of opportunity for "minorities" but also a program to give economic security to all? Searching for answers to these kinds of questions was at the heart of Coles's fieldwork, and it drew the respectful interest of the Nixon administration as he worked on *The Middle Americans*.[68] It also drew him toward political figures like Mondale whom, he hoped, would have the political skills and opportunity to build a new kind of New Deal coalition. The steps toward that goal were taken, he believed, by answering key questions such as, Why did such a large majority of whites oppose busing? Was it just racism and a knee-jerk resistance to having a difficult life unsettled further? Or did this "silent majority" have something more to tell us? When the desegregation challenge moved North, Coles moved with it. In 1968 and 1969 Coles observed busing from the ground up in thirteen cities outside the South.[69]

In *The Buses Rolled*, Coles documented how closely the protest against busing and other symbols of social change during the Second Reconstruction blended with surrounding grievances about a new sense of powerlessness and economic insecurity. Consider the following words from a white Boston mother who opposed crosstown busing:

> *I'm not against any individual child. I am not a racist . . . I just won't have my child bused to some god-awful slum school, and I don't want children from God knows where coming over here. We put our last cent into this home. We both work to keep up with the mortgage and the expenses. We're not rich; we can't afford to be generous at someone else's expense. We just want to live peacefully out here . . . We want to be able to go out of our house without carrying a gun or holding on to a German police dog. We want to see them off to school, and not*

sit for the rest of the day wondering, Are they safe, will there be a big fight, and are they afraid even to walk home if they miss the bus? That's why I'm opposed to kids from one kind of section to another; it's going against what's natural for both groups of people. And if they try to take our children across the city, just because some college professors and professional agitators . . . say they should go, we'll have a civil war right up here in the North! Anyway, no one asks us about anything, so I guess we have to learn how to band together and take care of ourselves.[70]

Coles follows this passage with a tough analysis of where perhaps he and his allies in the civil rights movement had missed something important:

For her and people like her all over the country school desegregation entails not only social and economic threats but a kind of political affront: others begin to flex their muscles and make their point, whereas "ordinary working people" . . . have to sit back and take it, always take it." Taking it, she registers surprise, annoyance, resentment, confusion, fear, anger—and, not least, a sense of powerlessness, conveyed sometimes in the briefest but most plaintive of questions: "Why us?" And, for a while, many of us involved in one way or another with the civil rights movement had little interest in answering such a question. She was a "racist" or a "covert racist"—that is, if we even thought about her at all. Chances are we didn't, at least not in the beginning . . . There were, in our minds, the blacks, a terribly hurt and denied people, and us, their allies.[71]

Placing the quotes by Coles and his subjects side by side in this section gives a clear sense of the socio-political challenge that liberals of the Great Society/Second Reconstruction period sought to master: building a strong coalition of the powerless, the migrant worker, the lower-middle-class and middle-class white men and women, pressed hard by rising property taxes and increasing uncertainty about the future, and newly enfranchised and energized blacks.

For black parents, the risks and fears were also great, but they were also mixed with the hope that present difficulties might bring substantial future dividends in terms of educational opportunity. In the following passage, Coles captures the difficult calculation of cost-benefit analysis that comes with participating on the front lines of social change:

"I try to tell my daughter every morning . . . that when she gets on that bus, she's building a future for herself, and when she enters that school, she's doing some more building, and when she leaves the school in the

> *afternoon, she's put in a day of building, besides learning her numbers and her letters." The child got the message. Often she seemed too serious, too intent on doing that "building" her mother kept on mentioning.*[72]

Some blacks worried about the special costs their children had to pay in the desegregation experiment. As one African American Berkeley woman shared, "We worry about the Negro children; is it too much for them to be at school with boys and girls who have so much more, are far better prepared?"[73] These new fears coexisted with older ones about the consequences of becoming *too* conscious of race:

> *"What does that white teacher mean asking me to do that? Why did the white girl say what she did to me?" Oh, I've spent a lifetime with those questions . . . I hope and pray . . . that this new busing program will help this generation of Negro children grow up differently.*[74]

Coles's experience with busing outside the South confirmed a finding of his work in the South: busing was generally hardest on the parents, not the children themselves. Children were beings of great capacity and adaptability. He wrote, "I never saw children get sick because they were being bused. I never saw children become emotionally disturbed because they were bused; I never saw children's school work suffer because they were bused."[75]

What was most important in determining the attitude of parent and child was where they believed they were headed as a result of participation in such an experiment. No matter how this question was answered, there was no model busing experience, no "City on a Hill," even in Berkeley, California, a city with a national reputation for being "radical." When he rode the school buses for one week in the fall of 1968, Coles found a community very much like the many others he would visit, where tension, anxiety and guarded optimism were all close neighbors. The Berkeley differences that did exist had to do with process: community members were actively brought into decision making at every point, emphasizing class as well as race issues, in a way that Coles deeply admired. In addition, the busing plan was a two-way process, which required sacrifice by both white and black and by rich and poor.[76]

The approach in Berkeley stood in positive contrast to the experience in Boston. In the fall of 1974, when busing came to Boston, Coles, in an interview with *Boston Globe* columnist Mike Barnicle, charged that the busing program was "a scandal . . . I do not think that busing should be imposed on working class people exclusively. It should cross these lines and people in the suburbs should share [in] it."[77]

Working in Public

When Coles advised Mondale in early 1972 about populist restlessness out in the land, he knew what he was talking about. Eight years later, as a Watergate-induced period of liberal reform was coming to an end, Coles called the decisive turn in American politics, when populism and conservatism became fused to create a long-term electoral majority.for the Republican Party, at least in presidential elections. Writing in April 1980 in the pages of the *New Republic*, Coles told of recent visits to the sites of his research during the 1960s and early 1970s. In the suburbs of Boston, the American South, and the working-class Chicano neighborhoods of Albuquerque, Coles observed the "politics of scarcity and fear at work: racial slurs directed at blacks for being demanding or 'lazy'; and at Indians for 'not being really being part of this country.'"[78]

To these Americans, Reagan possessed the honesty of President Jimmy Carter in combination with a toughness seen to be lacking in an overwhelmed incumbent. For his part, Carter was not the "fighting populist" in the mold of Harry Truman. Unlike Truman, who "stood his ground" even when public opinion was against him, Carter moved to where short-term political necessity directed him, a strategy that caused him, by the latter days of his term, to imitate rather than credibly distinguish himself from Reagan.[79] Consequently, Coles urged his readers not to underestimate the drawing power of Reagan, who, after all, had served as the governor of the nation's largest state for eight years without bringing disaster to state government.[80]

The economic and political disappointments of the previous decade had made the marginalization of Reagan as an out-of-touch extremist much more difficult than it had been in 1964 against Senator Barry Goldwater. In the sixteen years since the Johnson landslide that ushered in the Great Society, "the people . . . [have been] 'educated' from so many quarters, for so many recent years, to believe that liberalism is bankrupt, that any sudden attempt to paint Reagan into a reactionary corner may not be so easily managed."[81] The final scene was at hand: era of contention and reconstruction begun with *Brown* and the movement that had sparked that legal challenge was at an end; and Coles, who had been among its most astute participant-observers over twenty years, was there too, calling the turn right.

* * *

The reasons for the limited reach of the Second Reconstruction are most immediately rooted in the particular circumstances of the1960s and 1970s, but they extend back much further. The successful effort to short-circuit the Child Development Group of Mississippi in 1965 and 1966 should send us back, first thirty years to the New Deal–era battles over extending Social Security benefits to migrant and domestic workers (many of whom in the South were African American), and then back seventy years to the first Reconstruction and the unsuccessful struggle of former slaves to receive the promised "forty acres and a mule." Even in the best of times for the construction or elaboration of the American welfare state, this process has had to proceed within the strict limits imposed by history and political tradition, especially a vision of republicanism and of federalism that works against centralized action, especially if it is meant to expand the "welfare state."

When I first began thinking and writing about Robert Coles, I mistakenly thought of him as a social scientist. Although his knowledge of social science's core disciplines entitles him to several graduate degrees, Coles is a humanist in orientation with professional training as a physician. In his years in the American social field, he sought the truth of life not in survey data or controlled experiments but in placing his life beside other individual lives and offering to help, standing in one place long enough that his competence and motivations could be tested:

> *What I have tried to do is locate my body and mind where certain citizens are up against difficult times. . . . listening to them, watching them—and being watched by them . . . taking a long enough time to be confused, then absolutely certain and confidant, then not so sure but a little more aware of why one or another conclusion seems to be the best.*[82]

When I place Coles in historical context, I see a home front goodwill ambassador at work, a participant-observer in a movement that the pragmatic Fulbright did not support with his senatorial vote and international prestige. Coles was a representative from one segment of America (affluent, educated, and connected) to that much larger "other America" beyond his experience: those who were hard-pressed, oppressed, struggling, ignored, and poor and without power. As part of the process of interaction and exchange, these fellow citizens taught him to be wary of generalization in the name of advancing the theoretical good. Coles was someone who consciously and conscientiously straddled various "worlds" within American society so that the totality might be seen more clearly. In the final analysis, however, the political and social coalition of which Robert Coles was a leading member was simply not powerful enough to prevail.

Notes

1. I wish to thank Marisa Chappell, Hamilton Cravens, and Werner Sollors for their comments on an earlier draft of this essay. I am also grateful to an anonymous reviewer who encouraged me to develop the metaphor of "cultural exchange." This research could not have been completed without the unconditional cooperation of Robert Coles. I thank John M. Rhea and Larry Mastroni for their assistance in following up on my bibliographic leads and in proofreading. Without the indefatigable assistance of Benjamin Dettmar, I could not have navigated the unprocessed Coles papers at Michigan State University. Special thanks to Jade Chan for catching many errors in my text and ensuring clarity at key points. To the extent that fuzziness and other mistakes remain, they are mine alone.
2. Robert Coles, *Children of Crisis: A Study of Courage and Fear* (Boston: Little, Brown, 1967), xi.
3. At the start of their work in the South, Robert and Jane Coles had to rely on family resources. This work laid the groundwork for the support of the Field, New World and Ford Foundations. Jane Hallowell Coles was a twentieth-century member of the New England "gentry class" identified by Dorothy Ross. The Hallowells were New England bankers with a strong commitment to abolitionism. Coles's British-born father entered the upper class as an M.I.T.-trained engineer whose smart investments in Boston real estate during the Great Depression allowed him to set up trust funds for his two sons. Coles's mother was the Iowa-born daughter of a minister who knew the Bible exceedingly well and took great care in the moral and religious training of her children. Coles discussed the Coles/Hallowell family support for his initial research in a telephone interview with the author on July 27, 2007. On Coles's parents, see also Paul Wilkes, "Robert Coles: Doctor of Crisis," in *Conversations with Coles*, ed. Joy Woodruff and Sarah Carew Woodruff, 64–80 (Jackson: University of Mississippi Press, 1992); and Susan Hilligoss, *Robert Coles* (New York: Twayne Publishers, 1997), 1–2. On the founding generation of social science, see Dorothy Ross, *The Origins of American Social Science* (New York: Cambridge University Press, 1991), 53–55, 61–62.
4. The Children of Crisis series consists of the following volumes: *Children of Crisis: A Study of Courage and Fear* (Boston: Little, Brown, 1967); *Migrants, Sharecroppers and Mountaineers: Volume 2 of Children of Crisis* (Boston: Little, Brown, 1971); *The South Goes North: Volume 3 of Children of Crisis* (Boston: Little, Brown/Atlantic Monthly Press, 1971); *Eskimos, Chicanos, Indians: Volume 4 of Children of Crisis* (Boston: Little, Brown/Atlantic Monthly Press, 1977); and *Privileged Ones: The Well-Off and Rich in America: Volume 5 of Children of Crisis* (Boston: Little, Brown, 1977).
5. The phrase "America's Forgotten Children" appears on the cover of the February 14, 1972, issue of *Time*, with a photograph of Coles framed by children's drawings. A month earlier, *Time*'s closest competitor, *Newsweek*, devoted the entire religion section of its January 17 issue to a detailed and deeply admiring review of volumes 2 and 3 of Children of Crisis (see Kenneth L. Woodward, "A Good Neighbor," *Newsweek*, January 17, 1972, 81–82).
6. "Breaking the American Stereotypes," *Time*, February 14, 1972, 36 (emphasis added).

7. Norman H. Nye, Sidney Verba, and John R. Petrocik, *The Changing American Voter* (Cambridge, MA: Harvard University Press, 1979), 96–155. For a persuasive argument that civil rights activists led in the changing of public opinion, see Taeku Lee, *Mobilizing Public Opinion: Black Americans and Racial Attitudes* (Chicago: University of Chicago Press, 2002).
8. On the relationship between American society and the social sciences during the founding of these disciplines, see Mary O. Furner, *Advocacy and Objectivity in the Professionalization of Social Science* (Lexington: University of Kentucky Press, 1975); Thomas Haskell, *The Emergence of Professional Social Science: The American Social Science Association and Nineteenth-Century Crisis of Authority* (Urbana: University of Illinois Press, 1977); and Dorothy Ross, *Origins of American Social Science* (see n. 3). On the formative years of the social science in the United States, see also Martin Bulmer, *The Chicago School of Sociology: Institutionalization, Diversity and the Rise of Sociological Research* (Chicago: University of Chicago Press, 1984); Andrew W. Feffer, *The Chicago Pragmatists and American Progressivism* (Ithaca, NY: Cornell University Press, 1993); and Mark C. Smith, *Social Science in the Crucible: The American Debate over Objectivity and Purpose* (Durham, NC: Duke University Press, 1994). On the importance of psychology as a framework shared by experts and the general public for understanding public problems, see Ellen Herman, *The Romance of American Psychology: Political Culture in the Age of Experts* (Berkeley: University of California Press, 1996). For a perceptive sense of how the intellectual confinements present at the creation of the social science remain very much in force in our own time, see Alice O'Connor, *Poverty Knowledge: Social Science, Social Policy and the Poor in the Twentieth Century* (Princeton, NJ: Princeton University Press, 2001). On the material consequences of racial segregation for the development of the social sciences, see Jonathan Scott Holloway and Ben Keppel, "Introduction: Segregated Social Science and Its Legacy," in *Black Scholars on the Line: Segregation, Social Science and American Social Thought in the Twentieth Century*, ed. Holloway and Keppel, 1–37 (Notre Dame, IN: University of Notre Dame Press, 2007).
9. For a clear discussion of the consequences of this trend for social science, see Jeffrey Prager, Douglas Longshore, and Melvyn Seeman, "Introduction: The Desegregation Situation," in *School Desegregation Research: New Directions in Situational Analysis*, ed. Jeffrey Prager et al., 4–6 (New York: Plenum Press, 1986).
10. Hylan Lewis, "The Culture of Poverty Approach to Social Problems," reprinted in *Black Scholars on the Line*, 430 (see n. 8).
11. Ibid.
12. Henry Luce, "A Letter from the Publisher," *Time*, February 14, 1972, 2.
13. Michael Corbett, *Political Tolerance in America: Freedom and Equality in Public Attitudes* (New York: Longman, 1982), 65–85; and Mildred A. Schwartz, *Trends in White Attitudes Toward Negroes* (Chicago: National Opinion Research Center, 1967), 8–48. On the pattern of American public opinion toward the poor and the causes of poverty, see Joe R. Feagin, *Subordinating the Poor: Welfare and American Beliefs* (Englewood Cliffs, NJ: Prentice Hall, 1975). On the strong support for various specific forms of public assistance (including Aid to Families with Dependent Children) through the late 1980s as opposed to generalized opposition to "welfare," see Fay Lomax Cook and Edith J. Bennett, *Support for*

the *American Welfare State: The Views of Congress and the Public* (New York: Columbia University Press, 1992), 60–88. Between 1991 and 1995, the number of Americans who believed that "too much was being spent on welfare" rose from 40 percent to more than 60 percent, although R. Kent Weaver's careful study of public opinion and congressional action concludes that public opinion was not so forceful as to explain the election-year bipartisan effort to "end welfare as we know it" in 1996 [R. Kent Weaver, "Polls, Priming and the Politics of Welfare Reform," in *Navigating Public Opinion: Polls, Policy and the Future of American Democracy*, ed. Jeff Manza, Fay Lomax Cook, and Benjamin L. Page, 106–123 (New York: Oxford University Press, 2002)]. The most comprehensive one-volume discussion of changes in public opinion on key issues between the New Deal and Great Society years is Rita James Simon, *Public Opinion in America: 1936–1970* (Chicago: Rand McNally, 1974). The best synthetic analysis of how the politics of the Second Reconstruction has influenced American politics since is Thomas Edsall, *Chain Reaction: The Impact of Race Rights and Taxes on American Politics*, with Mary D. Edsall (New York: W. W. Norton, 1991). On the continuing prevalence of racial stereotypes in the American polity, see Richard A. Apostle, Charles V. Glock, and Marjean Suelze, *The Anatomy of Racial Attitudes* (Berkeley: University of California Press, 1983); and David O. Sears, Jim Sidanious, and Lawrence Bobo, *Racialized Politics* (Chicago: University of Chicago, 2000).

14. Robert Coles, interview with the author, Concord, MA, October 5, 2006.
15. Robert Coles to Roman Leverenz, 31 October 1983, Coles Papers, box 5, misc., University of North Carolina–Chapel Hill.
16. Robert Coles, *Doing Documentary Work* (New York: Oxford University Press, 1997), 60–61.
17. Ibid.
18. Ibid.
19. Robert Coles, *Study of Courage and Fear*, 3–4 (see n. 2).
20. Robert Coles, *Migrants, Sharecroppers and Mountaineers*, 25–26 (see n. 4).
21. Ibid., 26.
22. Ibid., 27.
23. Coles, *Study of Courage and Fear*, 43 (see n. 2).
24. Ibid., ix.
25. Robert Coles, "A Matter of Territory," *Journal of Social Issues* 20, no. 4 (October 1964): 44.
26. Ibid., 44–51.
27. Coles recounts his wife's criticism in an undated and unpublished essay titled "The American Character: Hopes, Hates, Binds," 6–7, Coles Papers, box 14, University of North Carolina–Chapel Hill.
28. Coles, *Study of Courage and Fear*, 33 (see n. 2); Oscar Lewis, *The Children of Sanchez: Autobiography of a Mexican Family* (New York: Random House, 1962); Allison Davis and John Dollard, *The Children of Bondage: Personality Development of Negro Youth in the Urban South* (Washington, DC: American Council on Education, 1940); John Dollard, *Caste and Class in a Southern Town* (New Haven, CT: Yale University, 1937); Abram Kardiner and Lionel Ovesey, *The Mark of Oppression: A Psychosocial Study of the American Negro* (New York: W. W. Norton, 1951); and Kenneth B. Clark, *Prejudice and Your Child* (Boston: Beacon Press, 1955). For Coles's published discussion of these works, see Robert

Coles and Joseph Brenner, "American Youth in a Social Struggle: The Mississippi Summer Project," *American Journal of Orthopsychiatry* 35, no. 5 (October 1965): 909–925.

29. Coles, interview with the author (see n. 14).
30. Coles, *Study of Courage and Fear*, 41, 46 (see n. 2).
31. Ibid., 47.
32. Ibid., 48–49.
33. Ibid., 48–49.
34. Ibid., 62.
35. Ibid., 49.
36. Robert Coles to David Riesman, 1 May 1967, Coles Papers, box 6, misc., University of North Carolina–Chapel Hill.
37. Coles to Moynihan, 29 September 1967 (see n. 15).
38. Coles, *Study of Courage and Fear*, 367 (see n. 2).
39. See Daniel Patrick Moynihan, "The President and the Negro: The Moment Lost," *Commentary*, February 1967, 31–45.
40. On the circumstances of the leak, see Robert Novak, *The Prince of Darkness: Fifty Years Reporting in Washington* (New York: Crown Forum Books, 2007), 134–35.
41. Earl Lewis, *In Their Own Interests: Race, Class and Power in Twentieth-Century Norfolk, Virginia* (Berkeley: University of California Press, 1991), 200–4. On deindustrialization in the urban core, see especially William Julius Wilson's *When Work Disappears: The World of the New Urban Poor* (New York: Alfred A. Knopf, 1996) and his *The Truly Disadvantaged: The Inner City, the Underclass and Public Policy* (Chicago: University of Chicago Press, 1987). On the political and intellectual context of the debate over the Moynihan report, see Holloway and Keppel, "Segregated Social Science," 18–20 (see n. 8).
42. Robert Coles, "The Observer and the Observed," in Coles, *Farewell to the South* (Boston: Little, Brown, 1972), 363–400.
43. Robert Coles, *The Middle Americans* (Boston: Little, Brown/Atlantic Monthly, 1971); and Coles, *The Buses Roll* (New York: W. W. Norton, 1974).
44. In 1986, when Coles revisited his New Orleans materials, he provided a longitudinal portrait of Hank, a New Orleans boy not discussed in the earlier Children of Crisis series. Hank's father was a "hard-drinking carpenter and house painter . . . militant segregationist, a constant heckler of Rudy and, not least, a member of the Ku Klux Klan." As a boy of seven, Hank thought that Rudy "had been 'crazy' for continuing her mob threatened education" but did not seem to share in his father's systematic racism (in this, Coles wrote, Hank took after his mother). Hank did not immediately adopt his father's attitudes, Coles suggested, because, "all during the 1960s . . . [Hank] had enough to do during those years simply dealing with the difficulties, dangers and demands of his family's emotional life." Within a few years, as Hank entered adolescence, however, he had adopted his father's combativeness, occasionally explosive temper, and racism. Robert Coles, *The Moral Life of Children* (New York: Atlantic Monthly Press, 1986), 37, 41–44.
45. Coles, *Study of Courage and Fear*, 52–53 (see n. 2).
46. Ibid., 53.
47. Ibid., 55.
48. Ibid., 57.

49. Ibid., 85.
50. Ibid., 83–84.
51. Ibid., x.
52. Ibid., xi–xii.
53. Coles, *South Goes North*, 561 (see n. 4).
54. Citizens' Crusade Against Poverty, *Final Report on the Child Development Group of Mississippi* (Citizens Crusade Against Poverty, 1965), 6. The Citizens' Crusade Against Poverty was organized by the allies of the Child Development Group of Mississippi in organized labor and allied groups in the civil rights community. Coles is one of thirteen members along with Kenneth Clark, A. Philip Randolph (founder of the Brotherhood of Sleeping Car Porters and an icon to the nonviolent civil rights movement of the 1950s and 1960s), and Paul Anthony, executive director of the Southern Regional Council.
55. Polly Greenberg, *The Devil Has Slippery Shoes: A Biased Biography of the Child Development Group of Mississippi* (London: MacMillan, 1969), 55–56.
56. Ibid., 34–35.
57. Coles, *South Goes North*, 561 (see n. 4).
58. Coles, *Eskimos, Chicanos, Indians*, xi–xii (see n. 4).
59. This comparison of Coles to Kennedy was inspired by Kenneth B. Clark's 1971 interview with Coles titled "Poverty Is Black and White," obtained through interlibrary loan from the William T. Boyce Library, Fullerton College, Fullerton, CA. In this interview, Coles describes his deep admiration for Kennedy as well as his encounter with Kennedy voters who became Wallace voters. Coles argues here that Wallace gives "cohesion to poor and middle-class white people."
60. George McGovern, who became the Democratic nominee for president in 1972, was a key figure in the congressional investigation of hunger in the United States as chair of the U.S. Senate Select Committee on Hunger and Human Needs. Walter Mondale, who was elected vice president of the United States in 1976 and the Democratic nominee for president in 1984, conducted hearings on the problems of migrant workers as chair of the Subcommittee on Migratory Labor of the U.S. Senate Committee on Labor and Public Welfare. McGovern and Mondale provided introductory remarks to Coles's *Uprooted Children: The Early Life of Migrant Farm Workers* (Pittsburgh, PA: University of Pittsburgh Press, 1970). U.S. senator Edward M. Kennedy, perhaps the most important liberal voice of those years (and an unsuccessful candidate for the Democratic nomination for president in 1980), provided an introduction to Coles's *Still Hungry in America* (New York: World, 1969). Kennedy's essay was to have been written by his late brother Robert. In a letter to Coles dated January 24, 1968, Robert Kennedy agreed to write the essay. Robert F. Kennedy to Robert Coles, 24 January 1968, Robert Coles Papers, box 14 (Kennedy Family Correspondence), Correspondence Series, Special Collections, Main Library, Michigan State University. Coles discussed his relationship with Robert Kennedy, including his agreement to write the forward, in a telephone interview with the author on July 14, 2008.
61. Robert Coles to Walter Mondale, 20 January 1972 (see n. 15).
62. In correspondence with his good friend Maria Piers, Coles expressed support for McGovern but feared that his fieldwork with more politically conservative families might be compromised. Coles to Piers, 12 July 1971 (see n. 36). In the late spring of 1972, with this work complete, Coles agreed to advise McGovern

on a variety of issues, including "some of the experiences I've had with blue collar workers and black people, the two groups of people that I guess he still has some difficulty in reaching, not that he should, in view of his magnificent voting record, and his real idealism of spirit." Coles to Piers, 22 May 1972 (see n. 36). On the "McGovern Community Health Committee," see Piers to Coles, 14 July 1972 (see n. 36).

63. Robert Coles, "Civil Rights Is a State of Mind," *New York Times Magazine*, May 7, 1967, 32. Coles's journalistic writing is collected in *Farewell to the South* (Boston: Little, Brown, 1972). Though not technically in the Children of Crisis series, the essays in this collection document his public thinking while he worked on volumes 1–3.
64. Coles, "Civil Rights Is a State of Mind," 36, 43 (see n. 63).
65. U.S. Senate, *Migrant and Seasonal Farmworker Powerlessness: Hearings of the Subcommittee on Migratory Labor of the Committee on Public Welfare*, 91st Cong., 1st and 2nd sess., July 28, 1969, part 2, 334. See also Robert Coles, "The Poor Don't Want to Be Middle-Class," *New York Times Magazine*, December 19, 1965, which gives substantial space to the social analysis provided by the very hardworking poor of their condition and aspirations.
66. *Migrant and Seasonal Farmworker Powerlessness*, 335 (see n. 65).
67. See James T. Patterson, *America's Struggle Against Poverty* (Cambridge, MA: Harvard University Press, 2000), 163–66; and Vincent J. Burke, *Nixon's Good Deed: Welfare Reform*, with Vee Burke (New York: Columbia University Press, 1974), 117–18.
68. In a letter to his editor, Peter Davison, Coles wrote, "I've recently been bothered for days in a row by the White House, which keeps trying to find out whether they can have the manuscript of the book." Robert Coles to Peter Davidson, 4 August 1970, Coles Papers, box 2, misc., University of North Carolina–Chapel Hill.
69. Coles, *Buses Roll*, 29 (see n. 43).
70. Ibid., 27–28.
71. Ibid., 38.
72. Ibid., 26–27.
73. Ibid., 36.
74. Ibid., 37.
75. Robert Coles, "Does Busing Harm Children?" *Inequality in Education* (March 1972), 25.
76. Coles, *Buses Roll*, 28–47 (see n. 43).
77. "Busing Puts Burden on Working Class, Black and White" [Mike Barnicle interview with Coles published in the *Boston Globe*], clipping in box 19, Coles Papers, University of North Carolina–Chapel Hill.
78. Robert Coles, "Ronald Reagan, Populist," *New Republic*, April 12, 1980, 16.
79. Ibid., 17.
80. Ibid., 16.
81. Ibid., 17.
82. Coles, *Study of Courage and Fear*, ix, vii–viii (see n. 2).

Genomics, Genetic Identity, and the Refiguration of "Race"*

Michael G. Kenny

Rapidly evolving genomic technologies are enabling the creation of new forms of biosocial identity and the realignment of old ones.[1] Evidence for this proposition appears regularly via the media. One example is a BBC story concerning "divided Lebanon's common genes." Dr. Pierre Zalloua, author of the study upon which this story was based, finds that many Lebanese Muslims and Christians share Phoenician ancestry as identified by certain markers on the Y chromosome—an issue disputed during Lebanon's civil war, when the rival sides claimed to be the "true" descendants of that ancient and honored people. Now it seems they both were right:[2]

> *I think it's a truly unifying message, and for me it's very gratifying. Lebanon has been hammered by so many divides, and now a piece of heritage has been unraveled in this project which reminds us that maybe we should forget about differences and pay attention to our common heritage.*[3]

But that which unifies can also divide. This is an outcome shaped by the political milieu, the moral commitments of the researcher, and the tools at hand. Is the emphasis on human similarity, difference, or unity within diversity?[4] The essays in this volume ring changes on that theme in historical perspective. Here, however, I am concerned with the present and future, the dynamics of a brave new world under construction, in which the Lebanese story is just one among many.[5]

Genomic assay technologies are capable of assessing the fine structure of individual genomes with increasing power and decreasing cost. They open a window to the deep past, the history of human populations in evolutionary time, revealing a stratigraphic record within the genome itself to be deciphered by those attuned to the temporal significance of

* I thank Prof. Paul Farber for the invitation to Oregon State University for participation in the 2006 conference on "Race, Science and Law" for which the initial draft of this paper was prepared. Thanks also to Peter Chow-White of Simon Fraser University's School of Communications and to GenomeBC for the opportunity to present a later version at a symposium on "Confronting 'Race': DNA and Diversity in the Digital Age" in May 2008.

base-pair sequences and variants. One leading technology, the DNA microarray (or "DNA chip"), was first introduced to the world in 1992 and has come to dominate large areas of medical genetic research as well as increasing use by personal ancestry search services.[6] These devices can be employed as "difference engines" capable of sorting people into "types" (or "human kinds") according to the investigator's purpose, whether it be the determination of relative disease risk, the search for ancient ancestral origins, or the identification of presently unknown members of one's genetic family.

A "haplotype" is such a kind, and is defined as "the combination of allelic states of a set of polymorphic markers" on a chromosome or chromosome region.[7] In simpler language, this definition is based on the fact that each individual of a species may differ from another in its combination of possible alleles at a given genetic sequence, and thus possesses its own individual "haplotype."

However, many markers cohere in groups that are inherited as a block, thus giving rise to "haplogroups" that reflect population history (e.g., migration, founder effects, demographic bottlenecks, natural selection, genetic drift). This is the sense in which "haplotype" will be used here, and the fact that humans fall into such groups has important implications for medical research, deep-time studies of genetic ancestry, and the future of race.

Haplotypes are, of course, *types*, but so are races, and the relationship between genetic and racial classification—and between expert and lay categories—is a complex, difficult, and ethically fraught issue, as the ongoing "race in medicine" debate amply illustrates.[8] This debate centers on the problem of whether conventional racial categories can function as surrogates for medically significant genetic difference. Some researchers have answered in the affirmative, while the opposition fears that such a use merely "reinscribes race in the language of genomics."[9]

Perhaps, but traditional races are themselves undergoing a process of redefinition, as shown by the launching of the International HapMap Project to outline common human haplotype variants by gathering and comparing samples from persons of northwestern European (from Utah in the United States), Japanese, Han Chinese, and Yoruban descent. The HapMap is currently in the process of expanding its coverage, and it can be expected that the internal diversity of the sample populations will become more apparent as time goes by, and essentialist approaches to race will become still more problematical.[10] One immediate result of this great venture has been the commercial development of microarrays based on the HapMap that can expeditiously determine individual haplogroup identity and hence provide a clue to that person's ancestral history as well as possible disease susceptibilities via gene association studies.[11]

However, it has been observed that "in the new genomic medicine, the uncritical use of racial and ethnic categories by those interested in biological difference often distorts the relationship between genetics, disease, and group difference" and thus that "models of human identity" need to be reexamined in light of the rise of new mapping technologies.[12] The HapMap Consortium was very attentive to some of the dangers involved in defining group difference—and specifically in defining just what kinds of "groups" were relevant to the task at hand. The relevant category was "population"—"a group of people with a shared ancestry and therefore a shared history and pattern of geographical migration." But the consortium also cautioned that "human populations are the products of countless social, historical and demographic processes, and therefore can never be sharply defined."[13]

> *Despite the many non-biological factors that contribute to population identities, the way that a population is labeled in the HapMap and described in publications will have implications for all members of the population, as all of them (and all members of closely related populations) might be affected by the interpretation and use of findings of future studies that use the HapMap.*[14]

Concepts such as population, race, and ethnic group easily elide one another in misleading ways, sometimes with unpleasant political consequences. An instructive example is the fate of the Human Genome Diversity Project (HGDP), which undertook a comparative survey of the genomes of indigenous tribal isolates "with the ultimate goal of understanding how and when patterns of diversity were formed. It also has the added benefit of providing information that is likely to prove useful to several areas of biomedical research."[15] But this endeavor precipitated a backlash from aboriginal advocacy groups that condemned the HGDP for scientific racism, colonialism, biopiracy, and worse.[16] The selection of any population for study is, of course, a strategic choice determined by the goals of the research—what it is that the genomic structure of a particular human grouping is supposed to illuminate. Tribes are reified groups of uncertain genetic significance. The same is true of races. Either way, there is ample scope for confusion between genetic and ethno-political identity.

As a case in point, a 2008 article in the *Lancet* describes research facilitated by the HapMap that used microarrays to identify genetic factors associated with osteoporotic bone fractures. The summary of the results declared that the "risk alleles . . . are present in more than one in five white people."[17] The "white people" in question are of British and Dutch ancestry, and the summary seems to imply—contrary to the warnings of the HapMap Consortium—that this conclusion applies to white people in general. But what exactly is a white person in genetic terms? Such loose phraseology would be unlikely to pass muster with regard to criteria that have been proposed for the use of the race concept in scientific and medical publications.[18] As one overview of the problem states, "A particular challenge for interdisciplinary teams will be designing their studies and reporting their results in ways that convey to the public the complexities of biological systems."[19]

In another case, a Scottish-based research team published microarray-based findings that seem to establish population-specific risk factors for colorectal cancer.[2020] The title of the paper is dryly professional: "Genome-wide Association Scan Identifies a Colorectal Cancer Susceptibility Locus on 11q23 and Replicates Risk Loci at 8q24 and 18q21."[21] Among the findings was the discovery that a specific gene "is associated with significantly different risk in Japanese compared to European samples."[22]

However, the headline of the BBC report on this study boiled its essence down to "Racial Clues in Bowel Cancer Find." The BBC quoted the lead researcher as saying that "this is the first time that a race-specific effect has been found for a genetic marker."[23] But the original article only states that the research team had turned up "the first evidence for a population-specific CRC susceptibility allele."[24] Somewhere here confusion lies, and with it the risk of covert essentialism that many have alluded to. In the process of translation to the public sphere, the clinical populations used in this study have become representatives of their so-called races—an outcome that, being mindful of the problems encountered by the HGDP, the founders of the HapMap sought to avoid.

And what of population admixture? Iceland's deCode Genetics, a company that specializes in disease-related genetic association studies, investigated an "ethnicity-specific risk of myocardial infarction" through a comparative study of Icelanders, Euro-Americans, and African Americans. It was discovered that the presence of a particular haplotype (HapK) conferred modest risk among the European-derived groups but significantly higher risk in African Americans.[25] HapK seemed certain to be of European origin and therefore must be present in African Americans through admixture, and in fact conferred somewhat different levels of risk in relation to degree of European ancestry, which was estimated through comparison with the Yoruba population assayed by the HapMap. In sum, the authors of this study claimed to have "found a strong correspondence between self-reported ethnicity and genetically estimated ancestry."[26] Of course, what they really meant by "ethnicity" is "race" in its common acceptation.

However, Africa has great genetic diversity due to the antiquity of its population, and it might well be supposed that the peoples of the diaspora share in that legacy.[27] In any event, the nature of the hypothesized causal relationship between HapK, or African ancestry, and risk of myocardial infarction remains unknown, and researchers can only suppose that there exists a "strong interaction between HapK and other genetic variants and/or non-genetic risk factors that are more common in African Americans."[28] Nevertheless, the authors of the deCode study conclude that these findings might provide the basis for the identification of "agents" that "may prove useful for primary or secondary prevention of heart attacks"—in other words an "ethnically" (or is it "racially"?) specific medication.[29]

The quest for genetic ancestry raises similar issues. One example is the rise of consumer-oriented services, such as the National Geographic Society's Genographic Project, of which the mentioned Lebanese study is a part.[30] Deploying a variety of technical means, these endeavors offer individuals a chance to situate themselves in relation to the history of humanity as a whole or to that of some specific ethnic/racial group. They create communities in the process of identifying them, and these communities may, in principle, involve many tens of millions of people or be very small indeed, ranging in scale from presently unknown living genetic relatives to global groupings reflective of the evolutionary history and migrations of the human species.

Medical genetics paved the way, with its findings in recent times facilitated by the emergence of the microarray and amplified by the World Wide Web. The identification of Mendelian disorders such as Tay-Sachs Disease with its Ashkenazi Jewish associations and of chromosome anomalies such as Down syndrome have led to the formation of communities of affliction among those suffering from them and their relatives and friends.[31] An interesting example of such community formation is discussed in Amy Harmon's Pulitzer Prize–winning *New York Times* series, "The DNA Age," a series that is itself an important aspect of the wider biosocial process that she documents.

In one case, parents of children with developmental difficulties learned of one another when it was discovered that their offspring shared a specific chromosome deletion (a condition only first identified as such in 2001, involving deletion of an entire chromosome segment).[32] One parent characterized the outcome as the creation of a kind of "tribe," and thus, as Harmon pointed out, "a genetic mutation became the foundation for a new form of kinship."[33] This new tribe took organizational form via the creation of the 22q13 Deletion Foundation, complete with annual meetings and T-shirts. As stated on its Web site:

> *The families who have children diagnosed with 22q13 Deletion Syndrome have much in common. To many, it may be obvious: children with special needs. But to us it goes much farther beyond that. Our families have their own special needs. It's the need to have the emotional support of other families that are going through the unique challenges we go through every day. You see, our children are in a very exclusive group. Our closest connection isn't at our children's school, or on the next block. It's hundreds, sometimes thousands, of miles away. That's why our Support Group is so vital to our children's future. There are currently around 380 members in our database worldwide (as of February 2008); this number includes children and adults.*[34]

The relation between haplotype and ethnic/racial identity becomes a directly salient consideration in genetic ancestry search services, of which there are now a number available that differ in specificity, cost, and marketing strategy. For example, though they are based on similar techniques—mitochondrial DNA and/or Y chromosome analysis—the Genographic Project and AfricanAncestry.com appeal to quite different motivations on the part of their clientele, as indicated by their respective statements of purpose.

The inexpensive and highly successful Genographic Project appeals to those attuned to the National Geographic Society's "family of man" unity-in-diversity ethos:

> *Where do you really come from? And how did you get to where you live today? DNA studies suggest that all humans today descend from a group of African ancestors who—about 60,000 years ago—began a remarkable journey. . . . In this unprecedented and real-time research effort, the Genographic Project is closing the gaps of what science knows today about humankind's ancient migration stories.*[35]

Participants are then invited to contribute a cell sample, and, in so doing, enhance the project's growing research database. Only mtDNA or the Y chromosome is in play here (the subscriber must choose which to have analyzed). The Y chromosome is inherited male to male down the ages, and mtDNA is inherited female to female, pointing back to the so-called Mother of all Mothers—the African Eve; both, however, slowly alter over time, thus giving rise to divergent but related haplotypes and hence a clue to population history. The project determines the haplogroup membership of a given Y chromosome or mtDNA profile and provides the client with a certificate of identity accompanied by a written description and map of the general migration pattern of that matrilineal or patrilineal group. As the project organizers note, these results can give no sense of the historical complexities embedded in the other 45 chromosomes. But I suspect the average subscriber believes that matrilineal and patrilineal lines are all that really matter in any event.

The client is also cautioned that though "you may be surprised to learn a new story about your genetic background, you will not receive a percentage breakdown of your genetic background by ethnicity, race, or geographic origin. Nor will you receive confirmation of an association with a particular tribe or ethnic group."[36]

AfricanAncestry.com has a somewhat different goal. It specifies for a largely African American clientele the likely African tribal origin of a particular Y chromosome and/or mtDNA haplotype.

> *Do you know where you are from? Where you are really from? Not the town you were born in or the city you call home. Before the Middle Passage and the time of slavery in the Americas. The place where your bloodline originated and the roots of your family tree truly began. Many of us know we come from Africa but could not claim a specific ancestral homeland. That is, until now!*
>
> *With African Ancestry, you can finally trace your maternal and/or paternal lineages to connect your ancestry to a specific country in Africa and often to a specific African ethnic group. Now you can know "where you are from" and redefine "who you are."*[37]

The establishment of African ethnic/racial identity is the centerpiece of the appeal, but, as with the Genographic Project, the technical means deployed by AfricanAncestry.com only permit a surmise about the geographical or ethnic origin of one's "PatriClan" or "MatriClan," as established through comparison with a proprietary African Lineages Database. Again, these procedures have nothing to say about the structure of the rest of the genome, and once more the Y chromosome paternal lines or mtDNA maternal lines are taken as the keys to personal identity. AfricanAncestry.com offers testimonials as to the powerful effects this new knowledge may have and provides a social networking service for its clients to share their impressions and experiences.[38]

Other search services are more ambitious with respect to the portion of the genome they survey. DNATribes.com, for example, expands coverage into the autosomal chromosomes via statistical comparisons of the client's single tandem repeat (STR) pattern with a proprietary world database. As the company points out, its methodology is a globally comparative variant of the forensic procedures used by police services for so-called DNA fingerprinting. Here, however, what are being looked for are matches with population-specific STR patterns.[39]

> *DNA Tribes® analysis identifies the peoples and places where your geographical genetic ancestry is most strongly represented. DNA Tribes is the only personal genetic analysis that compares your genetic profile to a population database that includes over 250,000 individuals from 801 populations around the world, including 577 indigenous populations.*[40]

The results are probabilistic and, if one reads the many caveats in the fine print, the conviction that you have somehow discovered whom you *really* are may be substantially diminished. DNATribes accordingly advises that it delivers "just the facts. The greatest danger in genetic analysis lies in over-interpretation of data. Genetic ancestry is complex and does not always correspond to familiar ways of thinking about ancestry."[41]

But it is far easier to think in familiar ways, even with regard to 23andMe, currently one of the most powerful and sophisticated genetic ancestry search services. 23andMe made its public debut in 2007, accompanied by considerable media buzz. Its findings are based on data produced by the Illumina HumanHap550 Chip, which, according to its provider "enables whole-genome genotyping of over 555,000 tagSNP markers derived from the International HapMap Project."[42] By this means, 23andMe assays the entire genome to generate data, identifying Mendelian traits such as blood-group antigens, risk factors for a large number of genetically correlated human ills, and information about ancestry based on a comparative global analysis of all forty-six chromosomes and mtDNA.[43] The results are displayed on chromosome maps that indicate in color the probable and sometimes different geographical origins of given segments of each diploid pair, accompanied by user-friendly animated diagrams of global and familial genetic similarity. Disease risk probabilities are also estimated, accompanied by caveats about the dangers of over interpreting the meaning of such knowledge.

As for ethnic/racial identity, in a supplementary technical white paper, 23andMe goes into the problem of "estimating genotype-specific incidence in the context of ethnic variation":

23andMe does not presume to know the non-genetic characteristics of the Customer. 23and Me assumes that ethnic similarity serves as a proxy for genetic and non-genetic similarity, and therefore provides the Customer, to the extent allowed by published research, with the option to choose among reference ethnic groups (U.S. census categories) when interpreting their genotypic results.[44]

Participants are also invited to share their genome with other 23andMe clients and compare it with theirs, as well as with that of various historical figures. The social aspect of this service feature has caught the eye of the media,[45] and, indeed, one of 23andMe's aims is to expand its endeavor into the Facebook world. One of the means by which it proposes to do this is by providing knowledge to its clients of personal Y chromosome and mtDNA haplogroups, with a nested history of these groups arranged as a family tree extending back into the deep past. Here, according to 23andMe, is what participants can do with this knowledge of their haplogroup:

1. *Find out who else in the world has it.*
2. *Find out how rare, or common, your haplogroup is.*
3. *Find out when a maternal (or paternal) ancestor of you and a friend (or significant other!) last walked the earth.*
4. *Find out how similar your haplogroup is to that of a friend or (in) famous person.*
5. *See if the geographic ancestry suggested by your haplogroup matches the information you have about family history.*
6. *Find other people with the same haplogroup and see if you can connect your family trees.*
7. *Find out what life was like in the time and place where your haplogroup originated.*
8. *Visit prehistoric sites inhabited by people with your haplogroup.*
9. *Post your maternal (or paternal) haplogroup on your Facebook page.*
10. *Get a vanity license plate with your haplogroup on it.*[46]

Given all this, it can be supposed that haplogroup membership will become an increasingly relevant aspect of genetically defined identities. What this implies for the future of race in admixed societies is for the future to determine, but it is at least of interest that 23andMe and other such services celebrated the much-delayed passage of the Genetic Information Nondiscrimination Act in the U.S. Senate, thus facilitating the sharing of genomic information via the Web.[47]

In an ethnographic spirit, I subscribed to 23andMe and learned that my paternal Y-haplogroup is R1b1c, which is a common northwest European variety. That makes sense, since family history indicates that my father's family is of English and Huguenot descent as far back as it can be traced. 23andMe provides a forum where I can fraternize with other R1b1cs and discuss our common past. Ancestry.com, another such service, points me to "RootsWeb, an Ancestry.com community" where I can do the same.[48]

According to 23andMe, I do not have the genotype for sickle cell disease, but I do have a 70 percent chance of having blue eyes (which indeed I do have), and I am a heterozygous carrier for one of the genes implicated in hereditary hemochromatosis, a fairly common iron-overload disorder overrepresented in persons of northern European descent—which a major medical study of this topic persists in calling "white people."[49]

My maternal identity proved to be rather a surprise, as did the hemochromatosis finding. According to 23andMe, I belong to mtDNA Haplogroup G, which is indigenous to Central Asia and found among Kazaks, Kyrgyz, and Tajik and appearing in varying frequency from the Urals to Japan. This is intriguing since my mother was a blue-eyed Swede; if the finding is correct, there are evidently complexities to the matrilineal side of my story that remain to be explored—perhaps the genetic memory of some atrocity committed by my imagined Viking forbearers! There is, so far as I can tell, no virtual Haplotype G community out there as yet, though many other haplotype-based communities are already to be found on the Web. So perhaps I should take the initiative and conjure up one of my own. Race, as we have seen, is a dynamic concept.

An unexpected new maternal identity makes an excellent conversation piece and, as such, is representative enough of the kinds of discussions these ancestry search services aim at provoking and encouraging. More significantly, enterprises such as AfricanAncestry.com are enabling the subdivision of the African diaspora into quasi-tribal "affiliative" communities based on supposed African tribe of origin.[50] This may be more problematical than had been realized but has no obvious political significance.[51] However, there are cases, such as the Lebanese or the Palestinian and Jewish Israeli, in which putative genetic identities can have significant repercussions when linked to ethno-political historical narratives and identity politics.

Consider the loaded title of a 2008 paper that appeared in the *American Journal of Human Genetics*: "The Genetic Legacy of Religious Diversity and Intolerance: Paternal Lineages of Christians, Jews, and Muslims in the Iberian Peninsula."[52] The author concluded that the high Iberian frequency of Y-haplotypes associated with Sephardic Jews expresses a history of voluntary or forced conversion, given the relative absence of that community in Spain today. This paper appears in the context of a debate about the nature of Muslim rule in Spain before the Christian reconquest in the fifteenth century, and that in turn articulates to the question of the status of Islam in Spain now, and in relation to the West at large since 9/11.

Another such example, of the many that might be chosen, pertains to the racial politics of Brazil, where a survey of mtDNA haplotypes "revealed an astonishingly high matrilineal contribution of Amerindian and Africans. Present-day Brazilians thus still carry the genetic input of the early colonization phase."[53] A wider survey of the genomics of the Brazilian population—"A Molecular Portrait of Brazil"—had a nation-building goal from the outset:

> *By proposing that the many white Brazilians, once aware that they have Amerindian and African mitochondrial DNA, would tend to value the genetic diversity of their own country more and would build a more just and harmonious society, the geneticists suggest parameters for identities and means for transforming social relations mediated by biology, more specifically by genomics.*[54]

One might also mention the metaphorical potential of the discovery, via Y-chromosome analysis, that Thomas Jefferson very likely had both a black and white family—but that is a long story in its own right.[55] Findings of this type thus carry a moral/political load related to "the authority and legitimacy of genetics in the definition of collective identities in the modern world."[56]

The genome contains a history, but first it must be read and interpreted and, as I said at the beginning, the outcome is shaped by the political milieu, the moral commitments of the researcher, and the tools at hand. Gene-sequencing machines and the DNA microarray are such tools, just as were the skull-measuring calipers of earlier generations of physical anthropologists. Behind these tools are the interpretive schemas—such as neo-Darwinian theory and population genetics—that guide research and shape its results. "Race" is most certainly being refigured. Old categories are undergoing a subtle shift and new ones are emerging that must be understood in terms of the complex motivations behind the classificatory impulse. This process has consequences "on the street" (as with personal ancestry search services) and on the conceptual level as a natural outcome of a slow drift in theory, technology, and political context. But how ideas about race, genomics, and collective identity will play out in the real world depends on factors that papers such as this can only anticipate rather than predict.

Notes

1. Paul Rabinow, "Artificiality and Enlightment: From Sociobiology to Biosociality," in *Incorporations*, ed. Jonathan Crary and Sanford Kwinter, 234–52 (New York: Zone Books, 1991).
2. Pierre A. Zalloua et al., "Identifying Genetic Traces of Historical Expansions: Phoenician Footprints in the Mediterranean," *American Journal of Human Genetics* 83, no.5 (2008): 633–42.
3. Natalia Antelava, "Divided Lebanon's Common Genes," http://news.bbc.co.uk/2/hi/middle-east/7791389.stm (accessed December 22, 2008; site now discontinued).
4. Though the Phoenician traits may be "commonly" shared, they are far from universally shared, since the study in question only identifies them in about one-third of its six thousand participants. And the results pertain only to men, since, of course, it is only they who can share and pass on the Y chromosome.
5. Dena S. Davis, "Genetic Research and Communal Narratives," *Hastings Center Report* 34, no. 4 (2004): 40–49.
6. Microarrays consist of known base-pair sequences affixed to some substrate, such as a glass slide or bead. The DNA sample to be assayed is chemically broken into bits and run in solution through the array; the bits of the sample that are "complementary" to a sequence in the array will bind to it and generate a signal, such as a flash of light when stimulated by a laser scanner. This process demonstrates what sequences the sample consists of and can point toward the structure of the sample genome as a whole.
7. Mark A. Jobling, Matthew Hurles, and Chris Tyler-Smith, *Human Evolutionary Genetics: Origins, Peoples and Disease* (New York: Garland Science, 2004).
8. Michael G. Kenny, "A Question of Blood, Race, and Politics," *Journal of the History of Medicine and Allied Sciences* 61, no. 4 (2006): 456–91; Nikolas Rose, *The Politics of Life Itself: Biomedicine, Power, and Subjectivity in the Twenty-first Century* (Princeton, NJ: Princeton University Press, 2006); and John H. Fujimura, Troy Duster, and Ramya Rajagopalan, "Race, Genetics, and Disease: Questions of Evidence, Matters of Consequence," *Social Studies of Science* 38, no. 5 (2008): 643–56. This controversy has generated a very extensive literature. For further details, see the bibliographies in Rose, *Politics of Life Itself* and the papers in *Social Studies of Science* 38, no. 5 (2008), a special issue devoted to the relationship between race, genomics, and biomedicine.
9. Troy Duster, "Race and Reification in Science," *Science* 307 (2005): 1050–51; and Nadia Abu El-Haj, "The Genetic Reinscription of Race," *Annual Reviews of Anthropology* 36 (2007): 283–300.
10. International HapMap Consortium, "A Second Generation Human Haplotype Map," *Nature* 449, no. 18 (2007): 851–61.
11. Examples are the Illumina HapMap 550 array and the Affymetrix GeneChip Human Mapping 500k Array. Given the known distribution of haplotype frequencies, the HapMap Consortium had determined that "researchers will be able to scan the entire genome by genotyping only approximately 500,000 tag SNPs, and not all 10 million common SNPs [single nucleotide polymorphisms]." Since the genome is organized into relatively coherent blocks of co-inherited genetic elements, certain representative elements may therefore stand for the entire block. International HapMap Consortium, "Integrating Ethics and Science in the International HapMap Project," *Nature Reviews/Genetics* 5 (2004): 467.

12. Sandra Soo-Jin Lee, Joanna Mountain, and Barbara A. Koenig, "The Meanings of 'Race' in the New Genomics: Implications for Health Disparities Research," *Yale Journal of Health Policy, Law, and Ethics* 1 (2005): 47.
13. International HapMap Consortium, "Integrating Ethics and Science," 469 (see n. 11).
14. Ibid., 471.
15. Luigi Luca Cavalli-Sforza, "The Human Genome Diversity Project: Past, Present and Future," *Nature Reviews Genetics* 6 (2005): 333.
16. Jenny Reardon, *Race to the Finish: Identity and Governance in an Age of Genomics* (Princeton, NJ: Princeton University Press, 2005).
17. J. Brent Richards et al., "Bone Mineral Density, Osteoporosis, and Osteoporotic Fractures: A Genome-Wide Association Study," *Lancet* 371 (2008): 1505.
18. See Judith B. Kaplan and Trude Bennett, "Use of Race and Ethnicity in Biomedical Publication," *Journal of the American Medical Association* 289, no. 20 (2003): 2709–16.
19. Race, Ethnicity, and Genetics Working Group, "The Use of Racial, Ethnic, and Ancestral Categories in Human Genetics Research," *American Journal of Human Genetics* 77, no. 4 (2005): 527.
20. Albert Tenesa et al., "Genome-Wide Association Scan Identifies a Colorectal Cancer Susceptibility Locus on 11q23 and Replicates Risk Loci at 8q24 and 18q21," *Nature Genetics* 40 (2008): 631–37.
21. The protocol for specifying chromosomal location (e.g., "11q23") is by chromosome number ("11"), chromosome section ("p" or "q," standing for short or long arms), and band ("23").
22. Tenesa et al., 635.
23. BBC News, "Racial Clues in Bowel Cancer Find," http://news.bbc.co.uk/2/hi/health/7319251.stm (accessed May 2, 2008).
24. Tenesa et al., "Genome-Wide Association Scan," 635 (see n. 20).
25. Anna Helgadottir et al., "A Variant of the Gene Encoding Leukotriene A4 Hydrolase Confers Ethnicity-Specific Risk of Myocardial Infarction," *Nature Genetics* 38, no.1 (2006): 68–74.
26. Ibid., 73.
27. Donald F. Conrad et al., "A Worldwide Survey of Haplotype Variation and Linkage Disequilibrium in the Human Genome," *Nature Genetics* 38, no. 11 (2006): 1251–60.
28. Helgadottir et al., "Variant of the Gene Encoding Leukotriene A4 Hydrolase," 71 (see n. 25).
29. See Jonathan Kahn, "How a Drug Becomes 'Ethnic': Law, Commerce, and the Production of Racial Categories in Medicine," *Yale Journal of Health Policy, Law, and Ethics* 4 (2004): 1–46; and Kahn, "Exploiting Race in Drug Development: BiDil's Interim Model of Pharmacogenomics," *Social Studies of Science* 38, no. 5 (2008): 737–58. See also Duana Fullwiley, "The Biological Construction of Race: 'Admixture' Technology and the New Genetic Medicine," *Social Studies of Science* 36 (2008): 695–735.
30. Like the HapMap, the Genographic Project has gone to great ethical pains to head off the suspicion that this endeavor is inherently exploitative of aboriginal populations.

31. The creation of genetically based biosocial communities has been greatly facilitated by the Online Mendelian Inheritance in Man (OMIM) database, which provides detailed technical accounts of specific genetic disorders and side links to testing services and support/advocacy groups. Mendelian Inheritance in Man, http://www.ncbi.nlm.nih.gov/sites/entrez?db=omim (accessed May 24, 2009).
32. The medical consequences of a 22q13 deletion are known as Phelan-McDermid Syndrome. Mendelian Inheritance in Man, http://www.ncbi.nlm.nih.gov/entrez/dispomim.cgi?id=606232 (accessed May 24, 2009).
33. Amy Harmon, "After DNA Diagnosis: 'Hello, 16p11.2. Are You Just Like Me?'" *New York Times*, December 28, 2007, http://www.nytimes.com/2007/12/28/health/research/28dna.html?scp=1&sq=After%20DNA%20Diagnosis&st=cse (accessed February 24, 2008). Strictly speaking, the problem here is not a "mutation" but a nonhereditary deletion of a chromosome segment.
34. 22q13 Deletion Foundation: Phelan-McDermid Syndrome, http://www.22q13.org/about.html (accessed April 30, 2008). Microarrays have been developed that can test all known chromosomal disorders at once (microarray-CGH–"comparative genomic hybridization"), thus expanding technical capacity for the creation and articulation of biosocial communities.
35. National Geographic Society, "The Genographic Project," http://www3.nationalgeographic.com/genographic/index.html (accessed May 2, 2008).
36. Genographic Project, "Frequently Asked Questions: Testing and Results," https://genographic.nationalgeographic.com/genographic/lan/en/faqs_results.html#Q16 (accessed January 5, 2009).
37. African Ancestry, Inc., "African Ancestry Is the Answer," http://www.africanancestry.com/aa-is-the-answer/index.html (accessed February 3, 2008). AfricanAncestry has a very sophisticated and evocative Web site.
38. "Over the years, our African Ancestry family members have shared with us the impact their results have had on their lives and how they have used their results. Some members have:
    - Shared their results with everyone in that lineage at their family reunion
    - Traveled back to their ancestral homeland with their families
    - Completed their family tree to pass down to future generations
    - Became involved in social causes in Africa
    - Reached out to African communities in the US
    - And much more"

    African Ancestry, Inc., "Once You Know Your Roots," http://www.africanancestry.com/once-you-know.html (accessed May 3, 2008).
39. Therefore, in principle, the DNATribes methodology could be used for ethnic profiling of otherwise unidentified forensic DNA samples.
40. DNATribes, "Population," http://www.dnatribes.com/populations.html (accessed May 4, 2008).
41. DNATribes, "Frequently Asked Questions," http://www.dnatribes.com/faq.html (accessed May 24, 2009).
42. The HapMap Consortium judges 555,000 markers to be a sufficient number of SNPs for assaying the entire human genome (see n. 11).
43. "We currently provide four kinds of ancestry analysis. One, Maternal Ancestry, traces your maternal lineage back through time from you to your mother, her mother, and all the way to the mother of all humans. Paternal Ancestry does the

same for your paternal lineage, tracing from you back to the father of all fathers. A third, Ancestry Painting, tells you where in the world each stretch along each of your 22 autosomal pairs is likely to have come from. The last, Global Similarity, assesses your relatedness to 10 regions that include more than 50 populations worldwide, as measured by the similarity of your DNA to people from those groups. With time, the number of populations we use for comparison purposes will grow." 23andMe, "Ancestry Features," http://www.23andme.com/ancestry/techniques (accessed May 4, 2009).

44. Joanna Mountain, Andro Hsu, Mike Macpherson, and Brian Naughton, "White Paper 23-02," 6, http://www.23andMe.com/for/scientists/ (accessed May 3, 2009). This paper cites Neil Risch's important paper in *Genome Biology* that argued that self-identified "race" can be a useful and reasonably robust surrogate for medically relevant genotypes, and proved to be a major catalyst of the "race in medicine" debate. Risch, "Categorization of Humans in Biomedical Research: Genes, Race and Disease," *Genome Biology* 3, no. 7 (2002): 1–12.
45. See, for example, Emily Singer, "Social Networking Hits the Genome," *Technology Review*, March 26, 2008, http://www.technologyreview.com/Biotech/20464 (accessed May 3, 2009).
46. 23andMe, "What Can I Do with My Haplogroup?" http://www.23andme.com/you/faqwin/maternaltopthings/ (accessed May 3, 2009).
47. Meredith Wadman, "Genetics Bill Cruises through Senate," *Nature* 453 (2008): 9.
48. Bryan Sykes, *The Seven Daughters of Eve* (London: Bantam, 2001); and Sykes, *Blood of the Isles: Exploring the Genetic Roots of Our Tribal History* (London: Bantam, 2006). Though many references overlap, a Google search for R1b1c generates approximately nine thousand hits. In *Blood of the Isles*, geneticist Bryan Sykes, founder of OxfordAncestry, a popular search service catering to people of British Isles descent, provides a popularized genomic history via analysis of paternal haplogroups. In *The Seven Daughters of Eve*, Sykes coins mythic names and romanticized historical itineraries for the ancestresses of the major British mtDNA lineages.
49. Paul C. Adams et al., "Hemochromatosis and Iron-Overload Screening in a Racially Diverse Population," *New England Journal of Medicine* 352, no. 17 (2005): 1769–78.
50. Alondra Nelson, "Genetic Genealogy Testing and the Pursuit of African Ancestry," *Social Studies of Science* 38, no. 5 (2008): 771.
51. Ann Gibbons, "Africans' Deep Genetic Roots Reveal Their Evolutionary History," *Science* 324 (2009): 575.
52. Susan M. Adams, "The Genetic Legacy of Religious Diversity and Intolerance: Paternal Lineages of Christians, Jews, and Muslims in the Iberian Peninsula," *American Journal of Human Genetics* 83 (2008): 725–36.
53. Juliana Alves-Silva et al., "The Ancestry of Brazilian mtDNA Lineages," *American Journal of Human Genetics* 67, no. 2 (2000): 458.
54. Ricardo Ventura Santos and Marcos Chor Maio, "Race, Genomics, Identities and Politics in Contemporary Brazil," *Critique of Anthropology* 24, no. 4 (2004): 371.
55. Annette Gordon-Reed, *The Hemingses of Monticello: An American Family* (New York: W. W. Norton, 2008).
56. Santos and Maio, "Race, Genomics, Identities and Politics," 370–71 (see n. 54).

Acknowledgments

This volume grew out of a conference held at Oregon State University on April 29, 2006. The title of the day-long meeting was "Race, Science & Law" and was part of the set of events sponsored by the History Department with funding from the Horning Endowment. The Horning Endowment was established by the will of the late Benjamin Horning in honor of his parents, Thomas Hart and Mary Jones Horning, and it supports two endowed chairs in OSU's Department of History as well as a large range of public events. The intent of the endowment is to help bridge the sciences and humanities, and this conference is typical of the activities it has sponsored.

The editors of this collection (and organizers of the conference), Paul Farber of Oregon State University and Hamilton Cravens of Iowa State University, are very appreciative of the generous funding from the endowment and for the administrative support from the Department of History at OSU. In particular, we would like to thank Ginny Domka and Christie van Laningham, who so professionally took care of all the conference details and arrangements. We also appreciate the support of the History Department faculty and Kay Schaffer, the Dean of the College of Liberal Arts.

During the conference, it became evident that the papers hung together in a coherent form—something always wished for in a conference but not to be taken for granted that it will happen! We wish to thank Mary Braun of the OSU Press for her help in guiding us through the process that has now resulted in what we hope is a useful and interesting volume. Outside reviewers and the OSU Press staff have contributed to helping us convert a set of conference papers into an edited volume of scholarly papers.

Paul Farber

Contributors' Notes

Hamilton Cravens is a professor of history at Iowa State University, where he teaches graduate and undergraduate courses in American history and the history of science and technology. He is the author or editor of several books, including *The Triumph of Evolution: The Heredity-Environment Controversy, 1900–1941*; *Ideas in America's Cultures: From Republic to Mass Society*; *Before Head Start: The Iowa Station and America's Children*; and *The Social Sciences Go to Washington: The Politics of Knowledge in the Postmodern Era*. He has two forthcoming works, *Great Depression: Peoples and Perspectives* and *Imaging the Good Society: The Social Sciences in the American Past and Present*. He is currently writing a synthetic history of ideas of race in modern American politics, science, and popular culture, and a brief history of methods and concepts in the modern social sciences. He has held three Distinguished Fulbright Chairs in Germany and the Netherlands, and has won fellowships and support from the Ford Foundation, the National Endowment for the Humanities, and the Hoover Institution and the Stanford Humanities Center at Stanford University.

Paul Farber is the OSU Distinguished Professor of History of Science Emeritus at Oregon State University. He is author of *Finding Order in Nature: The Naturalist Tradition from Linnaeus to E. O. Wilson*, and he is completing a book on the history of ideas on race mixing in the United States.

Melinda Gormley is a visiting assistant professor of Science, Technology, and Society at the University of Puget Sound. Her work focuses on twentieth-century American life sciences, assessing scientists' uses of their international networks and disciplinary expertise for political and social ends. She is currently writing a book on geneticist L. C. Dunn and his scientific community that analyzes American scientists' sociopolitical campaigns to counteract totalitarianism and discrimination.

John P. Jackson, Jr., is an associate professor in the Department of Communication at the University of Colorado at Boulder. His work focuses on the rhetoric and history of scientific arguments about race in the United States and, more generally, the interactions of the social and biological sciences about human nature. He is the author of *Social Scientists for Social Justice: Making the Case Against Segregation* and *Science for Segregation: Race, Law, and the Case Against Brown v. Board of Education*.

Michael G. Kenny is a professor of anthropology in the Department of Sociology and Anthropology at Simon Fraser University in Burnaby, British Columbia. In his earlier career, Prof. Kenny investigated oral historical traditions in the Lake Victoria region of Kenya, and then moved on to a social historical study of the concept of "multiple personality" in American psychological thought. This culminated in a monograph titled *The Passion of Ansel Bourne* that focused on biographical studies of paradigmatic cases through which this supposed disorder came to be defined. In more recent times, Prof. Kenny has researched aspects of the history of the eugenics movement with regard to racial issues and, by that route, became interested in the social implications of recent developments in genomics, a concern upon which his paper is this volume is based.

Ben Keppel is an intellectual and cultural historian with a particular emphasis on the issue of race and on the overlaps between popular culture and politics of various kinds. He is the author of *The Work of Democracy: Ralph Bunche, Kenneth B. Clark, Lorraine Hansberry and the Cultural Politics of Race* and the coeditor (with Jonathan Scott Holloway) of *Black Scholars on the Line: Race, Social Science and American Thought in the Twentieth Century*. The research on Robert Coles presented here is part of a larger study on the place of children in debates on poverty and race between the 1940s and the 1980s. Keppel is an associate professor of history at the University of Oklahoma.

Edward J. Larson is the author or coauthor of fifteen books, including *Sex, Race and Science: Eugenics in the Deep South* and *A Magnificent Catastrophe: The Tumultuous Election of 1800*. He is University Professor of History and holds the Darling Chair in Law at Pepperdine University.

Vassiliki Betty Smocovitis is a professor of the history of science in the Department of Biology and the Department of History at the University of Florida. She is the author of *Unifying Biology: The Evolutionary Synthesis and Evolutionary Biology*. Her interests are in the history of evolutionary biology and the history of the botanical sciences. She is working on a biography of George Ledyard Stebbins, Jr., a prominent figure in plant evolutionary biology.

Andrew S. Winston is a professor in the Department of Psychology at the University of Guelph in Ontario, Canada. His work centers on the history of racism and antisemitism in psychology and on the use of psychological research by racial extremists. He has also written on the history of the concepts of "experiment" and "cause" in psychology. He is the editor of *Defining Difference: Race and Racism in the History of Psychology*. From 2002 to 2008 he served as executive officer of Cheiron: The International Society for the History of Behavioral and Social Sciences.

Fay A. Yarbrough is an assistant professor of history at the University of Oklahoma. She is the author of *Race and the Cherokee Nation: Sovereignty in the Nineteenth Century* and has also published articles in the *Journal of Southern History* and the *Journal of Social History*. She is currently researching a project on Choctaw Indians and the American Civil War.

Index